Estuaries
A Physical Introduction
2nd edition

Keith R. Dyer
Institute of Marine Studies, University of Plymouth, UK

JOHN WILEY & SONS
Chichester · New York · Weinheim · Brisbane · Singapore · Toronto

First edition © 1973 John Wiley & Sons Ltd

Reprinted July 2000

Other Wiley Editorial Offices

John Wiley & Sons, Inc., 605 Third Avenue,
New York, NY 10158-0012, USA

WILEY-VCH Verlag GmbH, Pappelallee 3,
D-69469 Weinheim, Germany

Jacaranda Wiley Ltd, 33 Park Road, Milton,
Queensland 4064, Australia

John Wiley & Sons (Asia) Pte Ltd, 2 Clementi Loop #02-01,
Jin Xing Distripark, Singapore 129809

John Wiley & Sons (Canada) Ltd, 22 Worcester Road,
Rexdale, Ontario M9W 1L1, Canada

Library of Congress Cataloging-in-Publication Data

Dyer, K. R. (Keith R.)
    Estuaries : a physical introduction / Keith Dyer. — 2nd ed.
    p.   cm.
    Includes bibliographical references and index.
    ISBN 0-471-97470-6 (alk. paper). — ISBN 0-471-97471-4 (pbk. alk. paper)
    1. Estuaries. I. Title.
    GC97.D93 1997
    551.46'09 — dc21   97-14903
                                                                           CIP

British Library Cataloguing in Publication Data

A catalogue record for this book is available from the British Library

ISBN 0-471-97470-6 (hardback)
      0-471-97471-4 (paperback)

Typeset in 10/12pt Times from the author's disks by Dobbie Typesetting Limited, Tavistock, Devon
Printed and bound in Great Britain by Bookcraft (Bath) Ltd, Midsomer Norton, Avon
This book is printed on acid-free paper responsibly manufactured from sustainable forestation,
for which at least two trees are planted for each one used for paper production.

Contents

Preface

The first edition of this book was published in 1973 and fulfilled its purpose, as it has been widely used as an introductory text. When, after a gap of many years, I returned to teaching an undergraduate course on estuaries, I found that no replacement had been written. It was inevitable that very significant advances in understanding had been made in the meantime. As those advances were contained in an enormous number of papers in a variety of journals, they were not easily available to the students. Obviously, the time was right for an updated edition.

Reviewing the literature showed that much of the material in the first edition was still valid, but it was mainly focused on the tidally averaged situation in estuaries. Over the last 20 years the major advances relate to understanding the processes occurring within the tide. This is a logical progression, because one has to understand the general before one has the context into which the detail can be set, and conceptual models need to be formulated before the physics and mathematics can be brought in. The tidal averaged conditions considered rigorously in terms of equations of continuity of volume, mass and momentum established the foundations of mathematical modelling which developed to predict the effects of, for instance, effluent discharge, or topographic modification. The same processes were considered to operate during the tide, and the modelling started predicting what would happen within the tide. At the same time improved instrumentation gave the continuity of measurements and the spatial density of measurements that allowed refinement and adjustment of the models, making them more realistic. There was an iteration between models and measurements that gave a healthy progress in understanding. Models gave insights into what measurements needed to be done, or which processes needed to be quantified better, and the measurements refined the models. Now we understand better the relationship between the varying friction and the velocity and salinity profiles, we can define better the values of mixing coefficients that need to be used in different circumstances. However, there are many things we still need to do further work on. We still do not fully understand the relationship of the water flow and the density field in the three dimensions, how the vertical, longitudinal and lateral fields are interconnected, or how the flow relates to the topography. Turbulent mixing in the combined presence of density stratification and the sea bed is still an area of active research, and the least known part of the estuary is near the head, where changes of water depth and salinity during the tide are particularly dramatic, and sediment suspensions become important. Because of the effects of weather on the boundaries of the estuary, as well as internally, a stochastic approach to modelling coupled to the deterministic one is required. The current sea level rise is of great significance to estuaries, and the important aim will be to predict the changes in topography of the estuaries, and the effect there will be on effluent dispersal, water quality and environmental health. All challenges for the future.

In this book the emphasis is maintained on describing and explaining the physical processes that govern water circulation, mixing and salt and sediment transport. The large number of publications on estuarine modelling which rest on that understanding is only touched on. Consequently, there are many papers that the knowledgeable reader may find not quoted. The topic is getting to the stage where one textbook cannot cover the whole field in depth.

I would like to thank many people for help in the preparation of this book—the many that have inspired and excited me through the years in discussions about estuaries. Especially I would like to thank Pauline Framingham who has cheerfully and expertly helped with the preparation of the figures, the text formatting and general trouble-shooting, and Roy Lewis for very helpful comments.

Keith Dyer
Plymouth
October 1996

Notation

A	Cross-sectional area
$\overline{A}$	Mean cross-sectional area
A_m	Cross-sectional area at mouth
A_0	Tidal amplitude, amplitude of tidal fluctuational in cross-sectional area
A_s	Water surface area
$\boldsymbol{A}$	Tidal fluctuation of cross-sectional area
a	Amplitude of tidal height variation, a constant
a	Pipe radius
B	Surface area of interface
b	Estuary breadth
C	Concentration of a pollutant, tidal variation in suspended sediment concentration
C_0	Concentration of pollutant in segment at outfall
C_d	Drag coefficient
$\boldsymbol{C}$	Concentration of river water
c	A constant, suspended sediment concentration, interfacial wave propagation speed
$\boldsymbol{c}$	Tidal wave celerity
D	A depth
D	Thickness of salt wedge
$\boldsymbol{D}$	Dynamic depth
d	Layer thickness
E	Vertical exchange coefficient
F	Flushing number
$\boldsymbol{F}$	Flux of salt or pollutant through a cross-section
$\boldsymbol{F}$	Flushing rate
F_B	Internal Froude number

F_e	Estuarine Froude number
F_i	Interfacial Froude number
F_m	Densimetric Froude number
F_T	Barotropic Froude number
f	Fresh water fraction
f	Coriolis parameter, $f_1 = 2\omega\sin\varphi$, $f_2 = 2\omega\cos\varphi$
G	Rate of energy dissipation per unit mass of water
g	Gravitational acceleration
h	Estuary depth, water depth
J	Rate of gain of potential energy
K_x	Coefficient of longitudinal eddy diffusion
K_y	Coefficient of lateral eddy diffusion
K_z	Coefficient of vertical eddy diffusion
k	Friction coefficient
$\boldsymbol{k}$	Wave number $= 2\pi/\lambda$
k	Decay constant
L	Length of estuary, length of salt wedge
Ne	Estuary number
N_x	Coefficient of longitudinal eddy viscosity
N_y	Coefficient of lateral eddy viscosity
N_z	Coefficient of vertical eddy viscosity
n	A number
P	Ratio of surface to rms tidal velocity
P_x	Rate of transport of tidal energy across section at x
P	Rate of introduction of a pollutant
$\boldsymbol{P}$	Tidal prism volume
p	Pressure
$\boldsymbol{Q}$	Volume rate of salt or water transport per unit width
R	River discharge
Re	Reynolds number
R_f	Flux Richardson number
Ri	Gradient Richardson number
Ri_e	Estuarine Richardson number
Ri_L	Layer Richardson number
r	Exchange ratio
r	Radius of curvature of streamlines
S	Tidal fluctuation of salinity
S_0	Amplitude of tidal fluctuation in salinity
S_n	Mean salinity in segment n
S_s	Mean salinity of sea water
S_t	Stratification number
s	Salinity, subscripts, superscripts etc., the same as for u
T	Tidal period
$\boldsymbol{T}$	Flushing time

t	Time, time constant
U	Tidal variation of longitudinal velocity
U_0	Amplitude of tidal oscillation in longitudinal velocity
u_*	Friction velocity $= (\tau_0/\rho)^{0.5}$
u	Longitudinal velocity
u'	Turbulent velocity components
$\bar{u}$	Tidal mean or residual velocity
$\langle\bar{u}\rangle$	Tidal mean velocity averaged over depth
$\langle u \rangle$	Depth mean velocity
u_d	Deviation of longitudinal velocity from cross-sectional average
u_e	Entrainment velocity
u_f	Velocity of fresh water flow $= R/A$
u_g	Strength of gravitational circulation
u_s	Surface current velocity
u_t	Root mean square tidal current
u_T	Transverse deviation from cross-sectional average
u_V	Vertical deviation of longitudinal velocity from the depth mean
u_z	Velocity at depth z
V	Tidal variation of lateral velocity
V_f	Volume fresh water in a segment
V	Low tide volume of a segment of estuary
v	Lateral velocity, subscripts, superscripts etc., the same as for u
W	Tidal variation in vertical velocity
w	Vertical velocity, subscripts, superscripts etc., the same as for u
x	Longitudinal distance
y	Lateral distance
z	Vertical distance
z_0	Bed roughness length
α	Specific volume, proportion in tidal prism mixing
α	Coefficient in equation of state for salinity
δS	Surface to bottom difference in tidal mean salinity
δ	Equilibrium pycnocline thickness
ε	Molecular diffusion coefficient, fraction of ebb water not returning on flood
ζ	Elevation of water surface
ζ_m	Mean amplitude of tide
η	Dimensionless depth $= z/h$
θ	A phase angle, directional angle
κ	von Karman's constant $= 0.4$
λ	Wave length, a dimensionless estuary length
μ	Damping coefficient, coefficient of molecular viscosity
v	Diffusive fraction of upstream salt flux, kinematic viscosity
ξ_0	Amplitude of horizontal water movement

ρ	Water density
ρ_f	Density of fresh water
ρ_s	Density of surface water
σ_{tH}	Time of high water relative to that at the estuary head
σ_{rc}^2	Variance of dye distribution
τ	Shear stress
τ_0	Bed shear stress
φ	Dissipation constant
ϕ	Angle of latitude
ω	Angular frequency of tide, angular velocity of earth's rotation

Note: In this book the modern convention has been followed that salinity should be expressed without units, rather than as practical salinity units, psu; parts per thousand, ‰; or grams per kilogram, g kg^{-1}.

Chapter 1

Introduction

Estuaries are formed at the mouths of rivers, in the narrow boundary zone between the sea and the land, and their life is generally short. Their form and extent is constantly altered by the erosion and deposition of sediment and drastic effects are caused by a small rising or lowering of sea level. Estuaries are likely to be common during inundation, and rarer during retreat. These sea level alterations may be *eustatic*, variations in the volume of water in the oceans, or *isostatic*, variations in the level of the land. In the recent geological past there have been very large eustatic changes in sea level. About 18 000 years ago the sea level was about 100 m below its present level, the water being locked up in extensive continental ice sheets. As the ice retreated the sea rose at a rate of about 1 m a century, drowning the valleys incised by the rivers. This Flandrian transgression ended about 3000 BC when the sea level was more or less the same as at present. Since then some workers have suggested that minor fluctuations have occurred, but these are probably mainly isostatic in origin. Scotland is rising at a rate of about 3 mm yr^{-1} in response to the removal of the ice sheet, whereas areas formerly peripheral to the ice sheet, such as southern England and Holland, are now sinking at about 2 mm yr^{-1}. At the moment it is generally agreed that sea level is rising by about 1 mm yr^{-1} world-wide, and this may accelerate because of global warming, with some people predicting of a total rise of 1 m by the year 2100. Human influences are also producing local effects, particularly in areas of gas and oil extraction.

Larger eustatic changes are possible. If all the world's ice was melted, it has been estimated that the sea level would rise by about 70 m. If this happened, new estuaries would be formed in the upper valleys of present-day rivers. Little sediment would appear from the rivers, but large quantities would be available from renewed coastal erosion. A reduction of sea level would produce shallow estuaries which would quickly fill with sediment derived from the rejuvenated upper river valleys. In either case, because of the increased, or reduced, depths of the gulfs and seas into which the estuaries emptied, the tidal conditions would be modified. At present, following the Pleistocene ice age which overdeepened river valleys, and the subsequent inundation which flooded them, estuaries are both well developed and numerous. In geological terms this situation may not last long.

Though they are a particularly ephemeral feature of the earth's surface, estuaries have probably been extremely important in the world's development. They have generally high inflows of nutrients from the land, but, because of their large variability in conditions, tend to have a lesser diversity of life than other aquatic environments. Individual species are numerous, but they are specialized and often tolerant to large extremes of temperature and salinity. It is thought by many zoologists that estuaries may have been the most likely situation in which the first signs of organic life evolved. Almost certainly estuaries were the route by which, many millions of years later, animal life slowly adapted itself to a land-living and air-breathing existence. Estuaries are effective traps for sediment in their formative stages, but gradually fill until a balance is achieved between the sediment inputs, the shape of the estuary, its depth and the scouring of the currents. Thus a slow morphological development occurs. When the discharge of sediment by the rivers is higher than the coastal processes can disperse, deltas form. If the sediment discharge is lower, or the coastal processes active, the estuary can be a major indentation in the coastline, and there can be an equilibrium shape established by a balance between the sediment inputs and the physical processes. Much of the sediment present can have a marine rather than a terrestrial origin. A significant feature of most estuaries is a zone of high suspended sediment concentration near the head of the estuary, a turbidity maximum. This often contains high concentrations of contaminants to which are added pollutants from effluent discharges.

Estuaries have a wide range of forms, each of which has developed as an interaction between riverine and marine processes. Estuaries are continually evolving, changing their shape, adapting to changes in river flow and to weather patterns. Delays and modifications also occur within each estuary, so that each responds differently, with a different balance of magnitudes and time-scales. In some respects an estuary can be compared with an electronic filter, changing the phase and amplitude of the input signals so that the output has a different magnitude and timing. Determining the characteristics of the circuit, the processes that determine the impedances, resistances and capacitances is the interesting challenge of estuarine research. The range of physical time-scales is shown in Figure 1.1. However, their relative importances will vary with time in each estuary. Regular cyclic variations are produced by the semi-diurnal tidal variations in sea level, which may have significant diurnal contributions, and friction produces higher harmonics that distort the tidal curve and create asymmetry in the tidal currents. There are spring and neap cycles and seasonal variations. The currents produce turbulence and internal

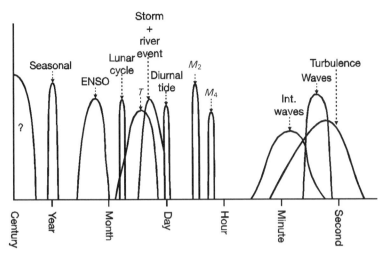

Figure 1.1 Diagrammatic representation of the physical time-scales affecting estuaries. T, Estuary flushing time; M_2, semi-diurnal tidal cycle; and M_4, quarter diurnal tide. The vertical energy scale is unquantified. Reprinted with permission from Dyer (1995a). Response of estuaries to climate change. In: *Climate Change Impact on Coastal Habitation* (Ed. D. Eisma). Copyright Lewis Publishers, an imprint of CRC Press, Boca Raton, Florida.

waves, create mixing, erode, transport and deposit sediment, and disperse pollutants. The seasonal variations are particularly evident in salinity as the estuary responds to river discharge. However, the salinity changes created by floods are affected by changes in vertical stratification and the time delay of the discharge impulse down the estuary. Weather events have time-scales of two to five days, which relate to the passage time for depressions, and the wind can affect the estuarine circulation, as well as creating waves and additional mixing. Thus it is obvious that the physical processes are complex and interacting, forming the driving force for many of the sedimentological, biological and chemical processes.

The circulation of water and mixing processes are driven by the density differences and the interaction between fresh and salt waters. Though there is only a density difference of about 2% between fresh and sea water, it is sufficient to create horizontal pressure gradients which affect the flow. The density of sea water depends on both the salinity and the temperature, but in estuaries the salinity range is large and the temperature range is generally small. Consequently, temperature has a relatively small influence on the density.

However, temperature can be a dominant factor at times. Surface heating can provide sufficient density differences between the estuary and the sea to maintain a gravitational circulation, i.e. the Spencer Gulf in South Australia. Because of the diurnal variation of temperature, these effects may be transitory in most estuaries. In many fjords there is virtually no river discharge in winter and surface cooling is intense. The surface waters can then become more dense than those at depth and will tend to sink. This vertical circulation phenomenon is known as *thermohaline convection*, and it creates a well-mixed water column. The effects of temperature, therefore, must not be forgotten.

Because of their fertile waters, sheltered anchorages and the navigational access they provide to a broad hinterland, estuaries have been the main centres of human development. The promotion of trade and industry has led to large-scale alterations of the natural balance within estuaries by alterations of their topography, making navigation for larger ships easier, and by large-scale pollution as a result of industrialization and population increases.

Deforestation of the land leads to increased run-off from the land, increased flashiness of the discharge and increased sediment load in the rivers. Building and paving of large areas also produce a quick response of run-off to rainfall. These effects may be controlled by building dams and may be reduced by the removal of river water for industrial processes and household use. However, maintenance of the river flow at a set level will decrease the natural tendency for rivers to flush sediments out of their estuaries and consequently may aggravate shoaling problems. Deepening of the estuary by dredging will increase the estuary volume and reclamation of intertidal areas will decrease the tidal flow, alter the mixing processes and circulation patterns and perhaps decrease the flushing time of the estuary. With a decreased flushing time the estuary cannot cope with and dispose of such large quantities of effluent. To understand and to be able to predict these effects is essential if humankind is not to do undue damage to the environment.

The main drawback in studying estuaries is that river flow, tidal range and sediment distribution are continually changing and this is exacerbated by the continually changing weather influences. Consequently, some estuaries may never really be steady-state systems; they may be trying to reach a balance they never achieve. Variability may be more important than the mean conditions. Because of the interaction of so many variables no two estuaries are alike and we must be careful to distinguish between unique details and general principles.

Many published studies of individual estuaries are available and there are several books that sift the details and produce the relevant general principles. These include Cameron and Pritchard (1963), Ippen (1966), Lauff (1967), Officer (1976), Kjerfve (1988) and Perillo (1995).

Chapter 2

Definition and Classification

DEFINITION

An estuary can be defined in a variety of ways depending on one's immediate point of view. Such definitions need to embrace the essential features and processes, as well as the context into which the estuary fits, and will lead into classification schemes. As far as most oceanographers, engineers and natural scientists are concerned, estuaries are areas of interaction between fresh and salt water, but there are over 40 different definitions of estuaries (Perillo, 1995).

The definition most commonly adopted is that of Cameron and Pritchard (1963), who state that 'An estuary is a semi-enclosed coastal body of water which has free connection to the open sea and within which sea water is measurably diluted with fresh water derived from land drainage'. This definition does not take into account the influence of the tide, which can be important even in the tidal reaches of the river above the saline intrusion. It does, however, hold for the tidally averaged situation.

The tidal effect has been emphasized by Dionne (1963): 'An estuary is an inlet of the sea, reaching into the river valley as far as the upper limit of tidal rise, usually being divisible into three sectors: a) a marine or lower estuary, in free connection with the open sea; b) a middle estuary, subject to strong salt and freshwater mixing; and c) an upper or fluvial estuary, characterised by fresh water but subject to daily tidal action'.

From the point of view of sediments, Dalrymple *et al.* (1992) have defined an estuary as: 'The seaward portion of a drowned valley system which receives sediment from both fluvial and marine sources and which contains facies influenced by tide, wave and fluvial processes. The estuary is considered to extend from the landward limit of tidal facies at its head to the seaward limit of coastal facies at its mouth'. This introduces waves as a significant process, but is restrictive of estuaries that do not receive marine sediment, even though they may have other estuarine features.

A further definition has been developed by Perillo (1995) as 'An estuary is a semi-enclosed coastal body of water that extends to the effective limit of tidal influence, within which sea water entering from one or more free connections with the open sea, or any other saline coastal body of water, is significantly diluted with fresh water derived from land drainage, and can sustain euryhaline biological species for either part or the whole of their life cycle'.

Possibly the most satisfactory overall definition would be an adaptation of the Pritchard definition as: 'An estuary is a semi-enclosed coastal body of water which has free connection to the open sea, extending into the river as far as the limit of tidal influence, and within which sea water is measurably diluted with fresh water derived from land drainage'.

CLASSIFICATION OF ESTUARIES

To compare different estuaries and to set up a framework of general principles, within which it may be possible to attempt a prediction of the characteristics of estuaries, a scheme of classification is required. Many different schemes are possible, depending on which criteria are used. Topography, river flow and tidal action must be important factors that influence the rate and extent of the mixing of salt and fresh water. Locally, and for short periods, wind also may become significant. The resultant mixing will be reflected in the density structure and the presence of stratification may cause modification of the circulation of water. Obviously all of these causes and effects are interlinked and it would be difficult to take account of them all in one classification system.

Pritchard (1952a) has separated estuaries into two groups: positive and negative. A positive estuary is one where the fresh water inflow derived from river discharge and precipitation exceeds the output from evaporation. Surface salinities are consequently lower within the estuary than in the sea. Negative estuaries are those where evaporation exceeds river flow plus precipitation, and hypersaline conditions exist, e.g. the Laguna Madre in Texas. Pritchard has called negative estuaries 'esteros', the Mexican term for these features, though this is not widely used.

Classification by Tides

The tidal context in which estuaries occur has been described in terms of the tidal range by Davies (1964):

Microtidal	< 2 m range
Mesotidal	< 4 m, >2 m
Macrotidal	< 6 m, >4 m
Hypertidal	> 6 m

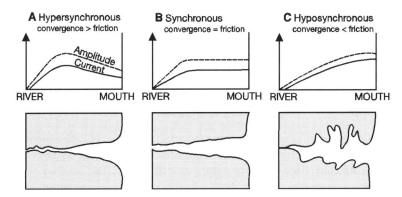

Figure 2.1 Diagrammatic representation of the modification of tidal range and current velocity amplitude in estuaries with varying geometries. Reproduced by permission of Springer-Verlag GmbH & Co. KG from Nichols and Biggs (1985: Figure 2.23).

In high tidal range estuaries the volume of water between high tide and low tide, *the tidal prism*, is large compared with the low tide volume. The interaction between the tidal wave propagating within the estuary and the morphology of the estuary leads to important variations in the range of the tide and the strength of the tidal currents. Convergence of the estuary sides causes the tidal wave to be compressed laterally, and, in the absence of friction, conservation of energy requires that the range of the tide increases. Friction in shallow water will cause a diminution in the tide range. Three conditions result, depending on the relative magnitudes of the two influences (Nichols and Biggs, 1985) (Figure 2.1).

1. *Hypersynchronous estuaries.* Convergence exceeds friction. As a consequence the tidal range and the tidal currents increase towards the head of the estuary, until, in the riverine section, the convergence diminishes, friction becomes the larger effect and the tide reduces. These estuaries are generally funnel shaped.
2. *Synchronous estuaries.* In these, friction and convergence have equal and opposite effects on the tide, and the range is constant along the estuary until the riverine section is reached.
3. *Hyposynchronous estuaries.* Friction exceeds the effects of convergence, and the tide range diminishes throughout the estuary. These estuaries tend to have restricted mouths, with the water entering through the mouth effectively spreading out within the estuary. The highest velocities occur at the mouth.

Classification by Topography

A topographic classification has been presented by Pritchard (1952b). He divides estuaries into three groups: coastal plain estuaries, fjords and bar-built estuaries.

Drowned River Valleys (Coastal Plain Estuaries)

These estuaries were formed during the Flandrian transgression by the flooding of previously incised valleys. Sedimentation has not kept pace with the inundation and the estuarine topography is still very much like that of a river valley. Consequently, maximum depths in these estuaries are seldom as much as 30 m. They have the cross-section of subaerial valleys and deepen and widen towards their mouths, which may be modified by spits. Their outline and cross-section are both often triangular. The width–depth ratio is usually large, though this depends on the type of rock in which the valley was cut. Extensive mudflats and saltings often occur and the central channel is often sinuous. The entire estuary is usually floored by varying thicknesses of recent sediment, often mud in the upper reaches, but becoming increasingly sandy towards the mouth. A remarkable characteristic of some is that the increase in cross-sectional area towards the mouth is exponential; this may reflect a long-term equilibrium adjustment between sedimentation and erosion by tidal currents.

Coastal plain estuaries are generally restricted to temperate latitudes, where, though river flow may be large at times, the amount of sediment discharged by the river is relatively small. River flow is generally small compared with the volume of the tidal prism (the volume between high and low water levels).

Examples: The Chesapeake Bay estuary system in the USA and the Thames, Southampton Water and Mersey in England.

Fjords

Fjords were formed in areas covered by Pleistocene ice sheets. The pressure of the ice overdeepened and widened the pre-existing river valleys, but left rock bars or sills in places, particularly at the fjord mouths and at the intersection of the fjords. These sills can be very shallow. In Norway a number have sill depths averaging 4 m and their presence can restrict the free exchange of water with the sea. The inlets of British Columbia, however, have deeper sills. Pickard (1956) has listed the physical features of several of these inlets and the sill depths are mainly between about 40 and 150 m. Inside the sills the maximum depth of the inlets reaches almost 800 m.

Because of overdeepening, fjords have a small width–depth ratio, steep sides and an almost rectangular cross-section. Their outline is also rectangular, but sharp, right-angled bends are common. Some fjords reach 100 km in length and the width–depth ratio is commonly 10:1.

Fjords generally have rocky floors, or very thin veneers of sediment, and deposition is generally restricted to the head of the fjord where the main rivers enter. River discharge is small compared with the total fjord volume, but, as many fjords have restricted tidal ranges within their mouths, the river flow is often large with respect to the tidal prism. Their occurrence is restricted to high latitudes in mountainous areas.

Examples: Loch Etive (Scotland), Sogne Fjord (Norway), Alberni Inlet (British Columbia) and Milford Sound (New Zealand).

Bar-built Estuaries

These estuaries could also be called drowned river valleys as they have experienced incision during the ice age and subsequent inundation. However, recent sedimentation has kept pace with the inundation and they have a characteristic bar across their mouths. This bar is normally the break-point bar formed where the waves break on the beach and for this to be well developed the tidal range must be restricted and large volumes of sediment available. Consequently, bar-built estuaries are generally associated with depositional coasts. The estuaries are generally only a few metres deep and often have extensive lagoons and shallow waterways just inside the mouth. Because of the restricted cross-sectional area current velocities can be high at the mouth, but in the wider parts further inland they rapidly diminish.

The river flow is large and seasonally variable and large volumes of sediment are riverborne at times of flood. The estuary form is governed by the river regime at the flood stage and may show a basin-bar structure caused by meander scouring. During the floods the bar may be swept completely away, but will quickly re-establish itself when the river flow diminishes. The mouth may undergo considerable variations in position from year to year. Variations of up to 3 km have been recorded for the mouth of the Vellar and the coastline around the estuary has built out by 300 m between 1931 and 1967.

Bar-built estuaries are generally found in tropical areas or in areas with active coastal deposition of sediments.

Examples: Vellar estuary (India), Roanoke River (USA).

Others

In this section we can include all estuaries that do not conveniently fit elsewhere. Included are tectonically produced estuaries: estuaries formed by faulting, landslides and volcanic eruptions.

Examples: San Francisco Bay, where the lower reaches of the Sacramento and San Joaquin Rivers have been drowned by movements on faults of the San Andreas fault system.

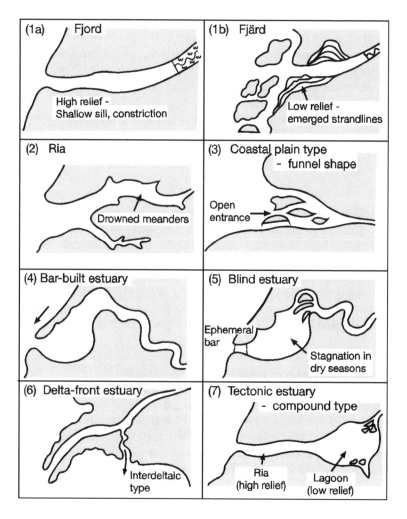

Figure 2.2   Estuarine physiographic types. Reproduced by permission of John Wiley & Sons Inc. from Fairbridge (1980).

Morphological Classification

Fairbridge (1980) has provided a comprehensive descriptive classification based on physiographic features. These are shown in Figure 2.2. The features are the result of the interplay of geomorphological history, river discharge of water and sediment, tidal currents and waves, and coastal processes. Within this classification, the types of Pritchard are major categories.

More recently Dalrymple *et al.* (1992) have considered the morphological development as part of an evolutionary sequence, which is determined by the

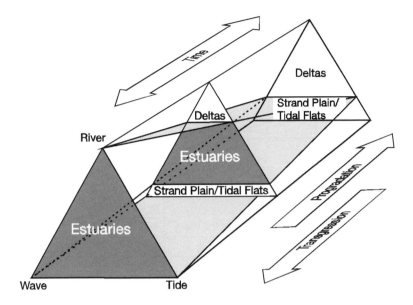

Figure 2.3 Evolutionary classification of coastal environments. Progradation requires a relative increase in sediment supply and sea level fall, transgression by a rise in sea level and reduction in sediment supply. Deltas form at the expense of estuaries with increased sediment supply. Reproduced by permission of Society for Sedimentary Geology from Dalrymple *et al.* (1992).

changing relative intensity of the river, wave and tidal influences. This can be represented by the triangular diagram shown in Figure 2.3, which is a convenient way of separating estuaries from deltas. Deltas are river valleys where the amount of sediment brought down by the river has more than kept up with the drowning of the coast by sea level rise. In situations where there is no river influence, tidal flats or strand plains will be formed. Estuaries occupy the centre of the triangle because they depend on all three influences. This is only a conceptual diagram because it is difficult to quantify the relative magnitudes. However, it is possible to extend the concept by introducing a time frame for a developmental sequence that would pertain under a progradational or a transgressive situation. During progradation the amounts of sediment coming from the rivers would increase and deltas would grow at the expense of estuaries. A similar diagram has been used to reveal the river, tide and wave influences on deltas, and as a means of classifying the morphology of deltas (Wright and Coleman, 1973). If the progradation is not depositional, but is due to a fall in sea level, then the tidal flats and strand plains are likely to grow because there will be overall reduced coastal relief and fewer valleys to be occupied by estuaries. During transgressions the drowning of the valleys will provide an increase in the number and extent of estuaries. If the sea level stays

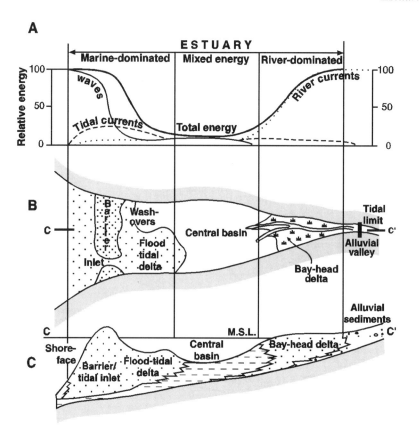

Figure 2.4 Distribution of (A) relative energy, (B) morphological components and (C) sedimentary facies in a longitudinal section within an idealized wave-dominated (microtidal) estuary. These estuaries are typically hyposynchronous, with the highest energy at the mouth. Reproduced by permission of Society for Sedimentary Geology from Dalrymple *et al.* (1992).

constant for a long period then it would be expected that many estuaries would largely fill in and reach an equilibrium form.

The interplay between the river, waves and tidal currents has been further considered by Dalrymple *et al.* (1992), especially in relation to the facies distributions. They consider two types: wave-dominated and tide-dominated.

1. *Wave-dominated* (Figure 2.4). In these estuaries the waves are significant at the mouth, where the sediment eroded from the coastline is transported alongshore to form a spit. This constricts the mouth and will build out until the tidal currents, which will gradually increase with the decreasing cross-section, achieve a balance in eroding sediment from the tip of the spit as fast

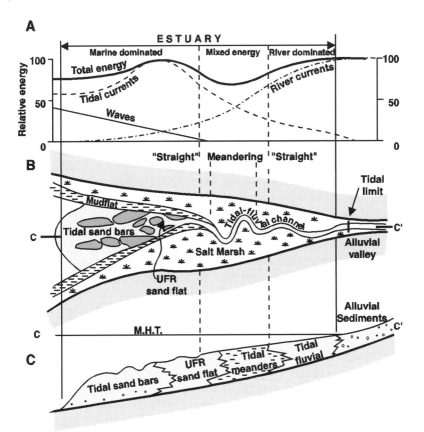

Figure 2.5 Distribution of (A) relative energy, (B) morphological components and (C) sedimentary facies in the longitudinal section for an idealized tide dominated estuary. UFR — Upper flow regime. These estuaries are typically macrotidal and hypersynchronous, with maximum energy towards the estuary head. Reproduced by permission of the Society for Sedimentary Geology from Dalrymple *et al.* (1992).

as it is brought there by the littoral drift. Within the estuary the tidal currents diminish rapidly. Towards the head of the estuary the relative strength of the fluvial processes increases. Thus there is a minimum in the total energy in the middle estuary where there are often extensive mudflats and marshes. In these terms energy is not strictly defined as it is related to the sediment transporting capability of the flows, which is due to a combination of the physical energy and the availability of sediment. Wave-dominated estuaries are likely to occur in micro- or meso-tidal conditions, and because the tidal currents diminish towards the head of the estuary they are likely to be hyposynchronous. An example is the Exe estuary in the UK.

2. *Tide-dominated* (Figure 2.5). As a result of the large tidal currents relative to the wave effects, the mouth area generally contains sandbanks which are

aligned with the current flow, and around which the sediment circulates. At the head of the estuary the tidal influence diminishes and the river flow becomes dominant. Here there are salt marshes which are often largely reclaimed land. These estuaries are likely to occur in macrotidal areas and in hypersynchronous conditions. An example is the Severn estuary in the UK.

Classification on Salinity Structure

As we have seen, there is a great deal of variety in estuaries caused by the differences in the tides, the river discharge, and the way these factors interact with the topography — a tide of a particular range is likely to produce different effects in a shallow estuary than in a deep one. The majority of estuaries that have been studied fall within the coastal plain category and it is apparent that within this group large differences occur in the circulation patterns, density stratification and mixing processes. Consequently, a better classification would be one based on the salinity distribution and flow characteristics within the estuary. Such a classification will lead to a better understanding of how the circulation of water in the estuaries is maintained, and quantification which should assist prediction. In the first instance we will consider the tidally averaged conditions; the intratidal conditions will be discussed in Chapters 8 and 9.

Pritchard (1955) and Cameron and Pritchard (1963) have classified estuaries by their stratification and the characteristics of their salinity distributions. They define four main estuarine types: highly stratified or salt wedge, fjords, partially mixed and homogeneous, or well mixed. The last group is subdivided into laterally inhomogeneous and sectionally homogeneous.

The Highly Stratified Estuary: Salt Wedge Type

Let us first consider an estuary emptying into a tideless sea, with a source of fresh water at its upper end. Let us also consider that the situation is frictionless, that the water behaves as a fluid without viscosity. Under these conditions the river water, being less dense than the sea water, flows outwards over the surface of the saline layer. The velocity in the surface layer, and its thickness, decrease towards the mouth as the estuary widens. The interface between the fresh and salt water, which is known as the *halocline*, would be horizontal and would extend up the estuary as a level surface. Because of the Coriolis force the seaward flowing river water would be concentrated on the right-hand side (looking downstream) in the northern hemisphere. There would be no mixing of salt and fresh water and no motion at all in the saline wedge. The velocity and salinity profiles would be as shown in Figure 2.6. The velocity would become zero at the upper surface of the salt wedge as defined by the salinity profiles, and would decrease towards the mouth as the estuary widened.

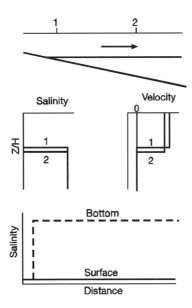

Figure 2.6 Salinity and velocity distribution and profiles in a frictionless estuary.

Let us introduce friction in the form of viscosity. There will now be a shear in the fluid flow near the interface which creates friction both on the salt wedge and on the fresh surface layer. The salt wedge will be pushed downstream until its upper surface has a slope sufficient to resist this force. The tip of the salt wedge will become blunted and the water surface will slope more steeply towards the sea. The Coriolis force will affect the lateral water slopes, with the interface sloping downwards towards the right and the sea surface downwards towards the left in the northern hemisphere.

Because of the velocity shear across the interface, a thin layer at the top surface of the salt wedge will be swept seawards. When the shear is sufficiently intense waves form and break on the interface and salty water is mixed into the surface fresh water. This process is called 'entrainment' (see Chapter 4) and is a one-way process. To preserve continuity a slight compensating landward flow is necessary in the salt wedge to replace the salt water passing into the upper layer. Thus the bottom water loses salt gradually into the surface layer and this loss is made good by a slow inflow of salt water from the sea. The entrainment adds volume to the water flow in the surface layer in its passage down the estuary and consequently the discharge increases towards the mouth. Typical velocity and salinity profiles are shown in Figure 2.7. The halocline is very sharp, but it is noticeable that the velocity falls to zero below the upper surface of the salt wedge, as defined by the maximum salinity gradient. Along the estuary salinity will be almost constant in both the surface and bottom layers,

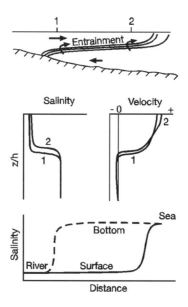

Figure 2.7   Salinity and velocity distribution and profiles in a salt wedge estuary.

except for zones at the tip of the wedge and where the halocline meets the surface. The surface gradient forms a distinct front that may be visible as a colour change in the water, and in many cases the front may be outside the mouth. The position of the salt wedge will vary with the river flow, and the tidal range is normally microtidal.

For this type of estuary, the ratio of river flow to tidal flow must be large and generally the ratio of width to depth is relatively small. The mouth of the Mississippi has a diurnal tide with a mean range of about 0.7 m and the river discharge is between 8.5×10^5 and 2.8×10^5 m^3 s^{-1}. In the Southwest Pass upstream flow prevails in the salt wedge regardless of the tidal phase, with downstream flow on the surface. This Pass is dredged to maintain adequate navigational access. In the shallower South Pass upstream flow occurs in the wedge during the flood tide at the same time as downstream flow occurs on the surface. During the ebb tide the flow in the wedge can be reversed and the current can be in a seaward direction at all depths (Wright, 1971).

Examples: Mississippi and Vellar estuaries.

Highly Stratified Estuary: Fjord Type

In many ways these estuaries are similar to the salt wedge type. However, the lower, almost isohaline layer is very deep. As river flow is dominant over tidal flow, entrainment is again the process mixing the fresh and salt waters. The thin upper layer is commonly of virtually constant thickness from head to mouth, but discharge again increases towards the mouth. In some fjords the upper layer thickness is restricted to a depth equal to the sill depth.

Where river discharge is high, the surface layer is almost homogeneous and the maximum salinity gradient occurs below the surface. Where run-off is lower, and near the fjord mouths, the surface layer is less homogeneous and the maximum salinity gradient occurs at the surface (Pickard, 1961). Though the temperature generally decreases with increasing depth, there are many instances where there are several marked maxima and minima. These occur especially in fjords with the inflow of melt water from glaciers (Pickard, 1971).

Because of the larger tidal velocities over the sills, mixing can be strong and the stratification weaker. The circulation over the sills may be entirely different from that occurring within the fjord. Generally, the compensating inflow over the sill is composed of a mixture of the coastal water and the outflow water. In the deeper parts of the fjord tidal action is small and there is often a layered structure showing successive intrusions of saline water. Often this renewal occurs annually and, occasionally, where the sill depths are small, renewal is so infrequent that anoxic conditions develop near the bottom. The saline inflow can also occur on a fortnightly basis with spring tides. It is best developed in the summer when the river flow is largest, entrainment is most active and the density difference between the deep fjord water and the coastal water is greatest. In winter, when the river discharge is low, surface cooling can produce thermohaline convection extending to the bottom. Salinity and velocity profiles for a typical fjord are shown in Figure 2.8.

Examples: Alberni Inlet (British Columbia), Silver Bay (Alaska).

Partially Mixed Estuary

If we now introduce tides into the estuary then the entire volume of the estuary will oscillate to and fro with the flood and ebb currents. It only requires a small tidal range to make this occur, though there is likely to be a particular range of tidal prism to estuary volume ratio over which it is effective in producing a partially mixed estuary.

The energy involved in these movements is large and it is mainly dissipated in the estuary by working against the frictional forces on the bottom, producing turbulence (see Chapter 4). The turbulent eddies lose their kinetic energy by working against the density gradients, thereby increasing the potential energy of the water column, and by viscous dissipation, creating heat. These eddies can mix both salt water upwards and fresh water downwards. Consequently, the salinity of the surface layer is considerably increased and to discharge a volume of fresh water equal to the river flow the seaward surface flow is enhanced. This causes an increase in the volume of the compensating landward flow. Consequently, a distinct two-layer flow system is developed in

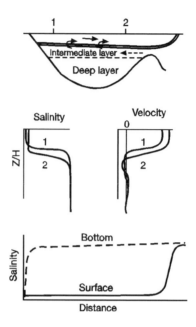

Figure 2.8 Salinity and velocity distribution and profiles in a fjord.

the mean flows. There is a height in the water column where the mean flow is zero, which is called the *level of no motion*, and this coincides with the level of the maximum vertical salinity gradient. Where the level of no motion reaches the estuary bed at the head of the salt intrusion there is a *null point*, where there is a convergence in the bottom velocities.

The process by which gravity, acting on horizontal density differences, creates a flow with dense water flowing beneath less dense water is termed *vertical gravitational circulation*. An essential part of this is mixing between the layers, because mixing tends to affect the longitudinal density gradients, which, in turn, alter the gravitational circulation.

Pritchard has calculated that in the James estuary the seaward flow in the upper layer is 20 times the river flow and the compensating inflow on the bottom is 19 times the river flow. It is now more difficult, however, to measure these flows because they are small relative to the oscillatory tidal flow superimposed on them. Some sort of averaging process is necessary to determine them (see Chapter 4), and we need to draw the distinction between the tidally averaged processes which reveal the residual flows, and those occurring during the tide.

Typical salinity and velocity profiles are shown in Figure 2.9. Because of the efficient exchange of salt and fresh water the salinity structure of the estuary is different from that of a salt wedge type. The surface salinity increases much more steadily down the estuary and undiluted fresh water only occurs very near

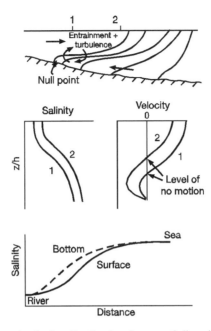

Figure 2.9 Salinity and velocity distribution in a partially mixed estuary.

the head of the estuary. In the saline water on the bottom there is also a longitudinal gradient of salinity. Consequently, there is a large section in the middle reaches of this type of estuary in which the horizontal salinity gradients are almost linear. The form of the vertical salinity profile does not change much along the estuary either. There is normally a zone of high salinity gradient at about mid-depth and the surface and bottom layers are almost homogeneous. The middle section has an almost constant surface to bed salinity difference. In the shallower parts on the estuary sides, however, the homogeneous bottom layer is missing and the maximum salinity gradient may occur near the bottom. In the upper layer surfaces with equal pressures slope towards the sea and in the bottom layer they slope towards the land. The lateral slopes are affected by the Coriolis force and are intensified as both outflowing and inflowing waters are deflected to opposite sides of the estuary.

The river flow must now be small compared with the tidal prism, and, as a consequence, most partially mixed estuaries are mesotidal. In Southampton Water, for instance, the maximum river flow is about $50 \, m^3 \, s^{-1}$, the maximum tidal flow is $7500 \, m^3 \, s^{-1}$, the tidal range varies between 5 and 1.5 m, and the tidal prism is about $10^8 \, m^3$.

In partially mixed estuaries the tidal range can change significantly between spring and neap tides. The spring tides increase the turbulent exchanges of salt and water, and as a consequence the vertical gravitational circulation can increase and the stratification can diminish. At times of high river flow the estuary will become more highly stratified and the intensity of the mean circulation should diminish. Within partially mixed estuaries there can be considerable variation of structure along the estuary, with highly stratified conditions near the head, where the water depth and the tidal range diminish, and river flow becomes comparatively more important. Additionally, well-mixed conditions can occur at the mouth where current velocities are higher.

Examples of partially mixed estuaries include the James River, the Mersey and Southampton Water.

The Vertically Homogeneous Estuary

When the tidal range is large relative to the water depth the turbulence produced by the velocity shear on the bottom may be large enough to mix the water column completely and make the estuary vertically homogeneous. It is difficult to be sure, however, that vertically homogeneous estuaries really exist, as small vertical variations may be lost in the averaging processes. Though there is a gravitational circulation produced by the denser water trying to flow landwards beneath the fresher water, it does not overcome the vertical mixing of the tidal currents and enhance the stratification. In these estuaries the tidal flow will be much larger than the river flow and this requires macrotidal conditions.

(a) Laterally Inhomogeneous.   When the estuary is sufficiently wide the Coriolis and centrifugal forces will cause a horizontal separation of the flow. According to the former, the seaward net flow will occur at all depths on the right-hand side in the northern hemisphere and the compensating landward flow on the left. This will either oppose or reinforce the effects of bends and cause the circulation to be in a horizontal plane rather than in the vertical sense as found in the other estuarine types. In wide estuaries this horizontal circulation can develop to the extent that there is a residual flow down the estuary on one side, and an inflow on the other, and a separation into flood and ebb dominated channels (Figure 2.10). The flood flow is stronger than the ebb in the flood channel and the water tends to move across the estuary to the ebb channel at about high water, with the reverse happening at low water. The central bank that separates the channels thus tends to be smaller than the tidal excursion of the water. The increase in salinity towards the mouth will be regular on both sides of the estuary (Figure 2.11). The lower reaches of the Delaware and Raritan estuaries are examples of this type.

(b) Laterally Homogeneous (Sectionally Homogeneous).   When the width is smaller, lateral shear may be sufficiently intense to create laterally homogeneous conditions. Salinity increases evenly towards the mouth and

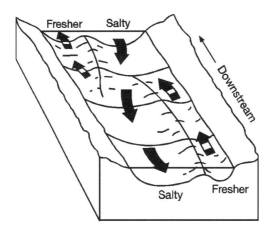

Figure 2.10 Distribution of tidally averaged flow in a well-mixed estuary, showing ebb and flood channels. Reprinted with permission from Dyer (1977) *Estuaries, Geophysics, and the Environment.* Courtesy of the National Academy of Sciences, Washington DC.

the mean flow is seawards throughout the cross-sections, tending to drive the salt out of the estuary. The balance is made by an upstream turbulent exchange of salt that is associated with the effect on the tidal flow of topographic irregularities and friction at the bottom. Salty water is trapped in embayments and near the estuary sides during the flooding tide and slowly bleeds back into the main body of water during the ebb stage. It is unlikely that this would be an effective mixing process unless the longitudinal salinity gradients are large. Consequently, the length of the estuary in which salinity is measurable may only be the equivalent of a few tidal excursions.

The estuarine type may show variations from section to section of the estuary. Near the head of the estuary where the tidal amplitude may be reduced, river flow can dominate, entrainment can be active and a highly stratified structure can result. Further downstream the tidal velocities will increase, turbulent mixing can become more active and a partially mixed structure will occur. Near the mouth the tidal currents may even be strong enough to produce well-mixed conditions.

Topographic differences are important factors influencing the flow structure in an estuary. Pritchard (1955) considers estuary depth and width to be important parameters in controlling an estuary's position in the sequence. If the river flow and the tidal range are kept constant and the estuary width is increased, the ratio of the tidal volume to river flow is changed, acting similarly to a relative reduction in the river flow. This will tend to cause a better mixed estuary. Similarly, increasing the depth will decrease the ratio of river flow to tidal flow, but the effectiveness of this will be offset by a decrease in the vertical tidal mixing. Thus the river flow becomes confined to the surface and the

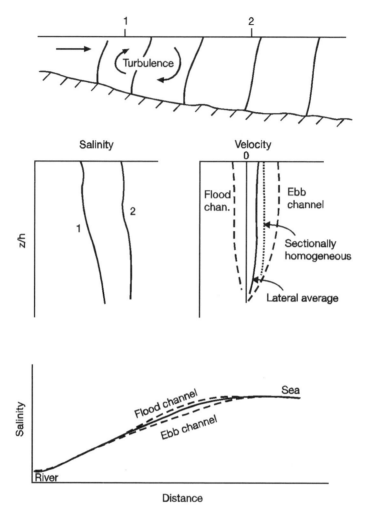

Figure 2.11 Salinity and velocity distribution and profiles in a well-mixed estuary.

estuary becomes more stratified. These effects are often shown near constrictions in estuaries, which because of the higher currents tend to be better mixed.

Quantified Classifications

There is obviously some correspondence between the results of classification on the basis of topography and salinity structure. It is clear, though, that the limits of each estuarine type are never well defined. The different types are merely stages on a continuous sequence. This sequence will be dominated to a certain extent by the river flow, which causes an inflow of buoyancy that tends to maintain stratification, and the tidal flow, which because of the friction causes mixing. Consequently, quantified classifications can be developed to characterise the ratio of these two factors, as well as the magnitude of the resulting salinity stratification. The major difficulty in this is that the gravitational circulation created by the density field modifies both the mixing and the stratification. Additionally, the tidal variation of water depth, or cross-sectional area, and the advection of the structure along the estuary needs to be included. There are many different schemes that have been developed, mainly from dimensionless numbers, or parameters, used with tidally averaged variables. We will now compare some of them as they illustrate the general principles and are useful comparitors for estuarine properties.

Simmons (1955) has found that when the *flow ratio* (the ratio of river flow per tidal cycle to the tidal prism) is 1.0 or greater, the estuary is highly stratified. When the flow ratio is about 0.25 the estuary is partially mixed and when it is less than 0.1 it is well mixed.

Alternatively, this definition can also be thought of as $P = u_f/u_t$, where $u_f = R/A$, R is the river flow, A the cross-sectional area, and u_t = rms tidal current. Then $P < 10^{-2}$ is well mixed, and $P > 10^{-1}$ stratified. Uncles *et al.* (1983) have used this for the Tamar estuary and show well-mixed conditions for the lower two-thirds of the estuary.

As the value of $u_t \approx 0.7U_0$, where U_0 is the amplitude of the tidal current, then stratified conditions will exist when $u_f/U_0 > 0.07$, and well mixed when $u_f/U_0 < 0.007$.

This criterion is a very general statement of the situation, as the estuary width and depth will have some control on the amount of mixing that a particular tidal rise and fall will produce in the water column. The Mersey and Southampton Water have flow ratios of 0.01–0.02 and yet are partially mixed, having vertical salinity differences commonly greater than 1.

An *estuary number* (Ne), can be defined as

$$Ne = \frac{P F_m^2}{T R} \tag{2.1}$$

where P = tidal prism volume; F_m = densimetric Froude number (Equation 2.5); T = tidal period and R = river discharge.

This is again an estuary-wide number which involves the ratio of the river discharge to the tidal flow, though it could be applied locally by using the tidal prism volume upstream of a point. When $Ne > 0.1$ the estuary is well mixed, and when $Ne < 0.1$ it is stratified.

Ippen and Harleman (1961) have considered the tidal energy and dissipation within estuaries. The rate of transport of tidal energy (P_x) across any section is

$$P_x = -cbg\rho A_0^2 \sinh(2\mu x)$$

where the wave velocity $c = \lambda/T$, T being the tidal period, λ the wavelength of the tidal wave, b is the breadth, ρ is the water density and A_0 is the tidal amplitude at the estuary head. The damping coefficient μ can be determined by using the curves on Figure 3.3.

For a purely standing wave without any progressive component, $\mu = 0$ and the total energy flux is zero.

The rate of energy dissipation in the portion of a channel between two cross-sections x_1 and x_2 is $P_{x_1} - P_{x_2}$ and the rate of energy dissipation per unit mass of water is

$$G = \frac{P_{x_1} - P_{x_2}}{\rho bh(x_1 - x_2)} \tag{2.2}$$

As a water particle moves down the estuary towards the sea, it gains potential energy due to its increasing density caused by mixing. The rate of gain of potential energy per unit mass over the entire length of the estuary L is

$$J = g\left(\frac{\Delta\rho}{\rho}\right)\frac{hu_f}{L} \tag{2.3}$$

where $\Delta\rho$ is the density difference between the fresh and ocean water and u_f is the mean velocity of the fresh water over the distance L. Thus, for a given estuary, J is affected only by variations in river discharge, and G indicates the amount of energy dissipated from the tide that is used either in mixing the water column, or liberated as heat. The ratio G/J, the *stratification number*, is a measure of the amount of energy lost by the tidal wave within the estuary relative to that used in mixing the water column. Thus it is rather like an efficiency.

The stratification number S_t is defined (Prandle, 1985) as

$$S_t = \frac{G}{J} = \frac{0.85kU_0L}{(\Delta\rho/\rho)gh^2u_f} \tag{2.4}$$

where k = friction coefficient $(= 0.0025)$; L = estuary length; U_0 = amplitude of the tidal currents; h = water depth; and u_f = depth mean current = R/A.

Values of $S_t < 100$ indicate stratified conditions, $100 < S_t < 400$ partially mixed, and $S_t > 400$ well-mixed conditions (Figure 2.12). Prandle (1985) shows by comparison with data that $\delta s/\langle s\rangle = 4S_t^{-0.55}$, where the angled brackets indicate a depth mean value, and δs is the surface to bed salinity difference. This means that $\delta s/\langle s\rangle < 0.15$ is well mixed and $\delta s/\langle s\rangle > 0.32$ is stratified.

If other factors are maintained constant, increasing river flow reduces the stratification number, indicating increased stratification. However, as the stratification number is also dependent on the length and depth of the estuary,

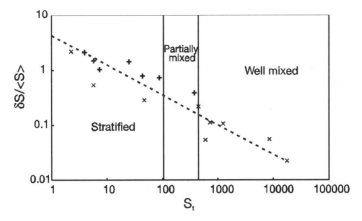

Figure 2.12 Stratification parameter versus stratification number S_t. ×, estuarine values, and +, flume tests. Reproduced by permission of Academic Press from Prandle (1985).

similar river discharges and tidal conditions in estuaries of different dimensions will produce different stratification numbers.

An example of this method, used successfully in the well-mixed coastal plain estuary of Coos Bay, is discussed by Blanton (1969). Its use in the Bay of Fundy and in the Delaware estuary is discussed in Ippen (1966).

Classification Using a Stratification–Circulation diagram

A further quantitative means of classifying and comparing estuaries, and one which requires measurements of salinity and velocity only, has been developed by Hansen and Rattray (1966). Further description of the analysis is made in Chapter 9. They have used two dimensionless parameters to characterize estuaries: a *stratification parameter* $\delta s/\langle s\rangle$, defined as the ratio of the surface to bottom difference in salinity (δs) divided by the average vertical or cross-sectional salinity $\langle s\rangle$, and a *circulation parameter* u_s/u_f, the ratio of the net surface current to the mean cross-sectional velocity. The circulation parameter expresses the ratio between a measure of the mean fresh water flow plus the flow of water mixed into it by entrainment, or eddy diffusion, to the river flow.

Their classification diagram is shown in Figure 2.13. In type 1 estuaries the net flow is seawards at all depths and upstream salt transport is by diffusion. Type 1a estuaries have slight stratification and coincide with the laterally homogeneous well-mixed estuary. In type 1b estuaries there is appreciable stratification, but no residual landward bottom flow. In type 2 estuaries the flow reverses at depth and corresponds to the partially mixed estuary. Both advection and diffusion contribute to the upstream salt flux. In type 3 estuaries the salt transfer is primarily advective. In type 3b estuaries the lower layer is so

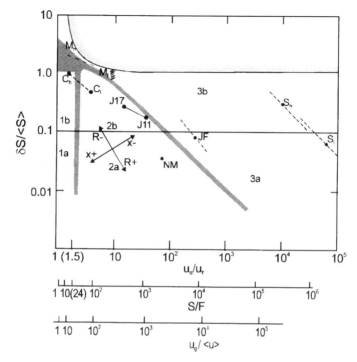

Figure 2.13 Hansen and Rattray's (1966) stratification–circulation diagram, together with alternative circulation scaling proposed by Prandle (1985) and Scott (1993). R+, R− shows direction of change with increasing/decreasing river flow. x+, x− shows direction of change for positions towards the mouth/head of the estuary.

deep that circulation does not extend to the bottom, e.g. fjords. Type 4 have more intense stratification: the salt wedge type. In Figure 2.13 the uppermost boundary represents conditions of fresh water outflow over a stagnant saline layer.

The advantage of the scheme is its relative simplicity, and it can be used to determine the proportions of total inward salt transport by gravitational circulation and by tidal dispersion. The disadvantages are that it uses tidal cycle means which are difficult to determine and which vary along the estuary. It does not include a finite amplitude tide, but use of the depth-averaged tidal mean Eulerian flow, rather than the river discharge, can take account of the additional Stokes' drift component.

The separation of the classes is arbitrary, and the data points from several estuaries show that estuaries are characterized by a line rather than a point. This arises because of two related effects: changes in river flow can cause both stratification and flow to change, and a point on the estuary to change its position on the diagram. Different sections in the estuary can have different positions on the diagram. Consequently, a change in river flow is equivalent in

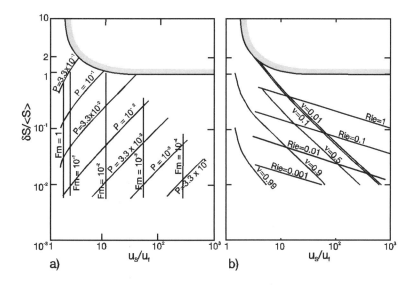

Figure 2.14 (a) Stratification–circulation diagram showing contours of the bulk parameters F_m and P. From Hansen and Rattray (1966). Reproduced with permission (b) Hansen and Rattray's stratification–circulation diagram with contours of dispersive fraction of upstream salt flux, and estuarine Richardson number Ri_e. After Fischer (1976) by permission of Annual Reviews Inc.

a general way to a change in position along the estuary. As the river flow increases there is a tendency for the salinity and mixing structure to be forced further down the estuary and vice versa. These effects are shown diagrammatically on Figure 2.13.

To relate the stratification and circulation to the forcing mechanisms, Hansen and Rattray present two further parameters which can be used to locate the position of an estuary on their diagram. These are the *densimetric Froude number*

$$F_m = \frac{u_f}{\sqrt{gh(\Delta\rho/\rho)}} \tag{2.5}$$

and the flow ratio $P = u_f/u_t$. $\Delta\rho$ is the ratio of the density difference between the sea water, which has a density ρ, and fresh water, and u_t is the root mean squared tidal velocity. They are shown on Figure 2.14. It is apparent that the circulation depends entirely on F_m, whereas the stratification depends on both. The diffusive (dispersive) fraction v of the upstream transport of salt is also related to these parameters (Figure 2.14b). This fraction is also discussed in Chapters 7 and 9.

Fischer (1972) introduced an *estuarine Richardson number* which represents the ratio of the gain of potential energy due to the freshwater discharge to the mixing power of the tidal currents. This is defined as

$$Ri_e = \frac{\Delta\rho}{\rho} \cdot \frac{gu_f}{bu_t^3} \tag{2.6}$$

where b is the estuary breadth. For an $Ri_e > 0.08$ the estuary is highly stratified, for $0.8 > Ri_e > 0.08$ it is partially mixed, and for $Ri_e < 0.08$ it is well mixed. This approach considers estuary breadth, rather than depth, as an important variable, the depth appearing on both the numerator and the denominator. Ri_e is related to the Hansen and Rattray parameters by $Ri_e = P^3/F_m^2$. The values of Ri_e plotted on the Hansen and Rattray diagram shows it to be strongly dependent on stratification, but only weakly dependent on circulation (Figure 2.14b). Thus, by replacing P $(= u_f/u_t)$ by Ri_e, Fischer showed that the stratification depends primarily on Ri_e.

This was confirmed by Oey (1984), who also came to the conclusion that Hansen and Rattray's analysis, which was specified for a prismatic channel, could be extended. Longitudinal variations in width, depth, river discharge, wind stress and mixing coefficients could be accepted, and the stratification–circulation diagram could be used with confidence under a wider range of conditions.

Prandle (1985) considered the density and the bed friction produced by the tidal motion, which generates the mixing, in terms of the dimensionless parameters:

$$S = \frac{h}{\rho} \cdot \frac{\delta\rho}{\delta x} \quad \text{and} \quad F = \frac{kU_0}{gh} \langle u \rangle$$

In his derivation he shows that the density forcing $S/2$ is partly balanced by a surface gradient of $-0.46S$, with the remaining component driving a residual circulation of $0.036S\langle u \rangle/F$ seawards at the surface, and $-0.029S\langle u \rangle/F$ landwards at the bed. This shows that the strength of the vertical gravitational circulation depends on the amplitude of the tidal currents, in agreement with the other derivations. Comparison with Equation 2.4 showed that

$$S_t = \frac{0.85}{\dfrac{S}{F}\left(\dfrac{\langle u \rangle}{U_0}\right)^2} \tag{2.7}$$

Thus the flow ratio parameter $(\langle u \rangle/U_0)$ must be introduced to determine stratification levels. Figure 2.15 shows a plot of S/F versus $\langle u \rangle/U_0$, which provides a comparison with the critical S_t values of Ippen and Harleman, and a means of calculating the vertical stratification for particular conditions.

The limits between the various categories of Hansen and Rattray have also been determined by Prandle (1985). The upper line marking the stratification

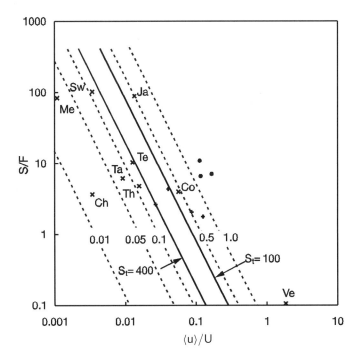

Figure 2.15 Classification diagram for **S/F** versus $\langle u \rangle / U$. Dashed lines show contours of $\delta s / \langle s \rangle$. Symbols show the values for different estuaries. Reproduced from Prandle (1985) by permission of Academic Press.

limit is shown to occur when the salty layer occupies almost the whole water depth in type 3(b) and less than one-quarter of the depth in type 4. The limit is

$$\frac{\delta s}{\langle s \rangle} = \frac{u_s / \langle u \rangle}{(u_s / \langle u \rangle - 1.26)} \tag{2.8}$$

The line separating types 1 and 2 depends on the absence of an upstream net flow at the bed. This occurs when **S/F** > 24, or $u_s / \langle u \rangle > 2$, which agrees with the Hansen and Rattray limit of 1.5 (Figure 2.13).

Jay and Smith (1988) analysed the flow and mixing in the Columbia River estuary and developed a classification scheme that considered the effect of a large tidal range in comparison to the water depth. They introduce a *barotropic Froude number* F_T, where $F_T = \zeta_M / h$, ζ_M being the mean tidal amplitude over the basin with mean depth h.

They also used an *internal Froude number* $F_B = d/D(\Delta \rho_h / \Delta \rho_v)^{-1/2}$. The ratio d/D is the ratio of the two layer thicknesses, $\Delta \rho_h$ is the horizontal density difference between the ends of the estuary, and $\Delta \rho_v$ is the vertical density difference in mid-estuary. This parameter ranges from near zero in inactive salt wedges, such as the Mississippi, to > 1 in weakly stratified estuaries.

They found that $F_B = 1$ separates highly stratified from weakly and partially mixed estuaries (Figure 2.16). Each estuary is shown by a zone within which the natural river flow variation occurs. The boundary between partially mixed and weakly stratified estuaries is the line of $F_T \sim F_B^{-1}$, this line being set by the Columbia River tidal amplitude. Though different estuaries may have different lines, it is argued that there are not likely to be significant differences. It is also thought that fjords are likely to occupy the extreme lower left of the diagram.

Scott (1993) derived three parameters representing discharge, salinity gradient and salinity that it is argued form a set from which many of the other discussed parameters can be derived. These three parameters can be plotted on the Hansen and Rattray diagram. He deduces that the circulation parameter $u_s/\langle u \rangle = 1.15 + 0.036 u_g/\langle u \rangle$. The value of u_g reflects the strength of the gravitational circulation and is defined as

$$u_g = \frac{gh^2 \alpha (\delta \langle s \rangle / \delta x)}{k U_0} \tag{2.9}$$

where α is the coefficient of salinity expansion in the equation of state. A scaling is provided for $u_g/\langle u \rangle$ as an alternative for the circulation parameter on the Hansen and Rattray diagram (see Figure 2.13). In this derivation, the stratification parameter S_t becomes

$$S_t = \frac{0.85}{\dfrac{u_g}{\langle u \rangle} \left(\dfrac{\langle u \rangle}{U_0} \right)^2} \tag{2.10}$$

The densimetric Froude number is also shown to have no dependence on horizontal and vertical mixing coefficients and thus to be an excellent estuarine discriminator. On the other hand, the estuarine Richardson number has a dependence on both coefficients.

There are also, of course, exceptions showing different characteristics to the generalized estuarine sequence. One well documented example is that of Baltimore Harbour, a small tributary of the Chesapeake Bay. There is very little river flow into the Harbour to contribute to the density stratification, but in Chesapeake Bay there is a normal stratification. Within the Baltimore Harbour tidal mixing processes are intense, creating higher surface salinity than in Chesapeake Bay and lower bottom salinity. At mid-depth the salinities are similar. This leads to a three-layer flow system with the inflow of fresher water on the surface, salty water on the bottom and outflow of intermediate density at mid-depth.

Similar three-layer flow effects have been found when strong winds blow up estuaries. The surface flow becomes reversed over the top few metres. Below this the outflow is considerably enhanced. The increased mixing tends to decrease the normal salinity stratification, but the normal pattern becomes re-established within a few days of the wind action ceasing.

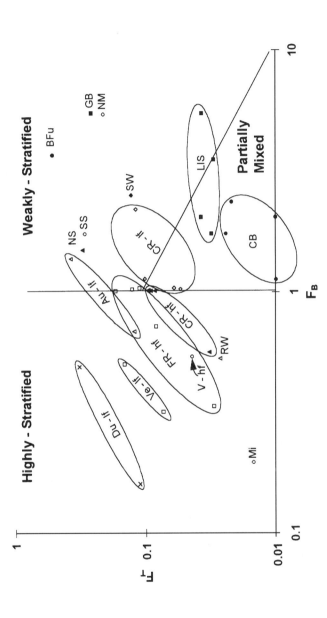

Figure 2.16 Classification system based on barotropic and baroclinic Froude numbers. lf, low flow; hf, high flow. Du, Duwamish; Ve, Vellar; Au, Aulne; Mi, Mississippi; Fr, Fraser; RW, Rotterdam Waterway; CR, Columbia river; CB, Chesapeake Bay; LIS, Long Island Sound; SW, Southampton Water; NS, North Santee; SS, South Santee; BFu, Bay of Fundy; GB, Great Bay; and NM, Narrows of the Mersey. Reproduced by permission of Springer-Verlag GmbH & Co. KG from Jay and Smith (1988).

In summary, it appears that the basic classification of Hansen and Rattray is a good means of comparing estuarine characteristics and of predicting the relative proportions of upstream salt transport carried on the gravitational circulation and by tidal diffusion. The various formulae can be used to calculate the magnitude of the various parameters. The circulation can be well represented by F_m. The stratification appears to be best represented by the estuarine Richardson number Ri_e. The main restriction is that the classification is for tidally averaged conditions, and gives no information concerning the processes that are related to the effects of a finite amplitude tide, or the variations within the tide. These will be considered in Chapter 9.

Chapter 3

Tides in Estuaries

As was made apparent in Chapter 2, there is considerable variation in the tidal conditions in estuaries and this has a major effect on the characteristics of an estuary. The shape of the tidal curve depends on the balance between the topography of the estuary and the effect that it has on the friction. The variation along the estuary of the amplitude and phase of the tide will affect the current velocities.

In an unstratified estuary the tidal wave will travel as a shallow water wave at a speed given by $c = (gh)^{1/2}$. If there is no friction, the wave entering a rectangular estuary will travel to the head where it will be reflected and return down the estuary. If the time taken to do this is equal to the tidal period, it will meet the next tidal wave entering from the sea. The reflected wave will then interfere with the wave just entering. A *standing wave* system can thus be set up in the estuary which will have an antinode at the head, where there will be a maximum in the tidal range equal to twice the amplitude of the original wave. At a distance down the estuary equal to one-quarter of the wavelength of the wave there will be a node with no tidal variation of water depth, but with a maximum in the horizontal currents. In short estuaries the node would be seawards of the mouth, the tidal amplitude would be at a maximum at the head, but the currents would increase towards the mouth. High and low waters would be simultaneous throughout the estuary and would coincide with the time of turn of the current. The tidal amplitude is thus 90° out of phase with the current velocity (Figure 3.1a). Also the salinity variation would reverse at high and low water. In longer estuaries several nodes and antinodes can be present.

If the energy of the tidal wave is completely dissipated by friction before reflection, or if the channel is infinitely long, then the tidal wave becomes solely a *progressive wave* in nature. The amplitude of the tide and the magnitude of the tidal currents diminish towards the head of the estuary and there is a

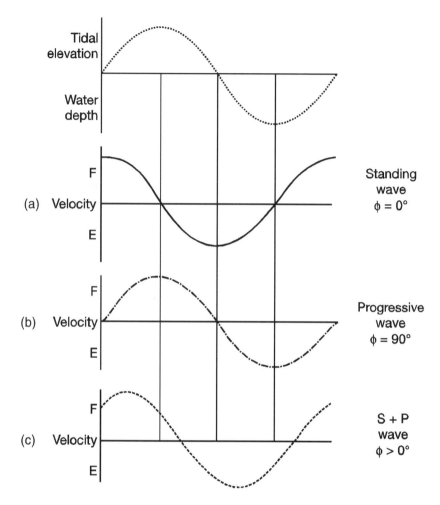

Figure 3.1 Idealized tidal response in estuaries showing the velocity variation for a standing tidal wave, a progressive wave, and a combination of the two types.

progression in the times of high and low water and the turn of the current along the estuary. In this case the tidal amplitude and the current would be in phase, i.e. the maximum flood currents would occur at high water (Figure 3.1b).

The relationship between the three variables elevation, velocity and salinity, can be shown by plotting them against each other. Figure 3.2 shows a graph of tidal height against velocity. The plot is an open ellipse for a standing wave, whereas it is a line for a progressive wave. In fact, most estuaries have some dissipation of the tidal energy before and after reflection and the tidal response

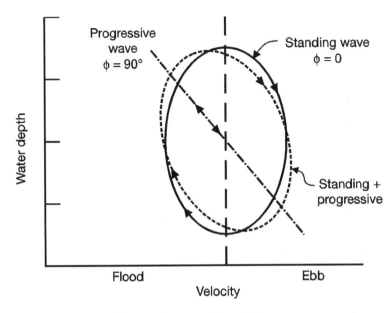

Figure 3.2 Hysteresis diagram for a standing tidal wave, a progressive wave and a combination of the two types.

is a mixture of a standing wave with a progressive contribution of variable magnitude. This means that at most positions the current continues flooding until after the elevation starts to recede (Figure 3.1c), and this provides the water for the elevation further upstream to continue to rise. The progressive wave contribution can be estimated by measuring the time difference between high and low water and slack water. This can then be expressed as a proportion of the tidal period in degrees. The maximum salinity would occur at about the end of the flooding current for both situations, i.e. it would be maximum at about high water for the standing wave, but at mean tide level for the progressive wave. However, there are delays in the time of the salinity maximum due to density current effects and cross-channel processes. Though these phase shifts are generally of the order 5–10°, they are an important feature in the transport of salt within the estuary.

Depending on the relative magnitudes of the standing wave and progressive wave components, the tidal amplitude and the timing of high and low water and slack tide vary along the estuary. At Cuxhaven on the Elbe, for instance, the flood and ebb currents start about one and a half hours after low and high water, and the time of high water propagates up the estuary at about $6 \, \mathrm{m \, s^{-1}}$. The tidal amplitude gradually diminishes towards the head of the estuary.

Ippen and Harleman (1961) have developed expressions that relate the relative times of high water along the estuary $\sigma_{\tau H}$, and the tidal amplitudes, to

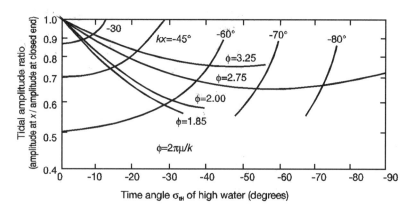

Figure 3.3 Nomogram for the determination of μ and k from tidal amplitudes and time of high water. From Ippen and Harleman (1961). Reproduced with permission.

the phase change kx (the wave number $k = 2\pi/\lambda$, λ being the wavelength of the tidal wave), and a damping coefficient μ which specifies the change in amplitude with distance x along the estuary caused by friction. The time of high water at any position in the estuary relative to that at the head $\sigma_{\tau H}$ is given by:

$$\tan\sigma_{\mathrm{tH}} = -\tan(\boldsymbol{k}x)\cdot\tanh(\mu x) \tag{3.1}$$

In a channel of uniform cross-section and roughness, $\boldsymbol{k}$ and μ would be expected to be constants and $\mu = (\varphi/2\pi)\boldsymbol{k}$, where φ is a dissipation constant. The values of μ, φ and $\boldsymbol{k}$ can be determined for any estuary by using the nomogram on Figure 3.3, with measured values of the time angle of high water and the relative tidal amplitudes measured at a number of distances x from the head of the estuary.

In the shallow water of estuaries two processes affect the tidal wave. The first is that even in a frictionless estuary, when the tidal variation of the water depth is large, the wave crest will move more quickly than the trough. The crest of the tide may partially overtake the trough, resulting in a shorter flood and a longer ebb, and the highest velocities thereby occur on the flood tide. The second is the effect of bottom friction. This is a non-linear process which depends on the square of the current and its effect is to produce greater friction in shallow water. This also slows down the water movement at around low water relative to high water. Thus the time delay between low water at the mouth and that at the head is greater than the time delay of the high water. The combined effects of these two processes produces a short duration flood phase of the tide, and fast flood currents, creating what is known as *flood dominance*. In some estuaries the asymmetry can become marked and a bore can develop, with the tide coming in abruptly as a visible wave.

Ebb dominance can also be produced within estuaries essentially by interactions between the deep channels and the shallow water areas, and the varying distribution of friction during the tide. This effect has been considered, among others, by Boon and Byrne (1981), Fitzgerald and Nummedal (1983) and Friedrichs and Aubrey (1988). If an estuary has a cross-sectional area at the mouth of A_m, and a water surface area A_s, the velocity through the mouth u must be given by $A_s \cdot \mathrm{d}h/\mathrm{d}t = A_m \cdot u$. Around high water there is a large surface area relative to the cross-sectional area, and A_m/A_s is a minimum. Conversely, at low water A_m/A_s is at its maximum. Thus the water surface within the estuary adjusts much more rapidly to the tide at the mouth at low water than it does at high water. It can be said that the estuary is more efficient in its exchange of water at low tide than at high tide (Fitzgerald and Nummedal, 1983). Consequently, the lag in elevation between the ocean levels and those in the estuary are less at low water than at high tide, and this implies a steepening of the tidal curve within the estuary during the ebbing tide, so producing a faster ebb current. This effect can be further enhanced by large changes in the cross-sectional area at the mouth during the tide (Boon and Byrne, 1981).

The effect of frictional distortion of the tidal curve can be considered in terms of tidal harmonic analysis, and the production from the M_2 tide of the M_4, M_6, etc overtides. The major part of the asymmetry of the tide curve can be represented by superposition of M_2 and M_4, both in terms of height and velocity

$$A = a_{M2} \cos(\omega t - \theta_{M2}) + a_{M4} \cos(2\omega t - \theta_{M4})$$
$$u = u_{M2} \cos(\omega t - \varphi_{M2}) + u_{M4} \cos(2\omega t - \varphi_{M4})$$

(3.2)

where a and θ are the amplitude and phase of tidal height, and u and φ are the amplitude and phase of the tidal velocity. The elevation phase of M_4 relative to M_2 is $2M_2 - M_4 = 2\theta_{M2} - \theta_{M4}$. The elevation amplitude ratio is $M_4/M_2 = a_{M4}/a_{M2}$. Speer and Aubrey (1985) have considered the implications of these ratios and show that for an elevation phase ($2M_2 - M_4$) between 0 and 180° the system will be flood dominant, and for a phase of 180–360° it will be ebb dominant. In either case the larger the M_4/M_2 ratio, the more distorted and the more strongly flood or ebb dominated the system becomes. Thus the dominance can be predicted from tidal analysis.

Friedrichs and Aubrey (1988) use a mathematical model and a comparison with field data to quantify the causes for the differences. The model has two distinct elements: a deep channel that transports the momentum and broad tidal flats that store water. Their results confirm that in flood dominant systems the channels are responsible for frictional distortion, whereas in ebb dominant systems the storage of water on the tidal flats is greatest. They find that two non-dimensional parameters can be used to represent these characteristics. The ratio a/h, the ratio of the M_2 amplitude at the mouth to the mean estuarine water depth, represents the relative shallowness of the estuary. This is the same

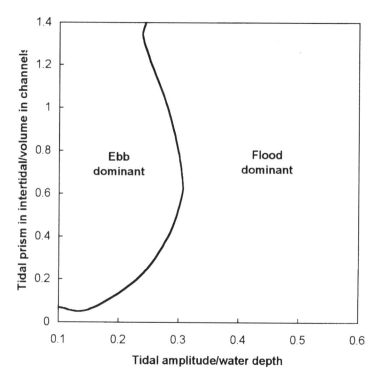

Figure 3.4 Ebb and flood dominance in estuaries related to their volumetric and tidal characteristics. After Friedrichs and Aubrey, 1988.

as the barotropic Froude number of Jay and Smith (1988). The ratio V_s/V_c, the ratio of the volume of water over the tidal flats to that in the channels, is a measure of the relative amounts of water stored on the tidal flats. Their results are shown in Figure 3.4. This diagram is a useful way of estimating whether an estuary is flood or ebb dominant from simple morphological information.

It can be seen by comparison with the descriptions in Chapter 2 that flood dominated estuaries tend to be macrotidal and hypersynchronous. Ebb dominated estuaries tend to be microtidal and hyposynchronous. Synchronous estuaries are likely to be neither flood nor ebb dominant.

EQUILIBRIUM ESTUARIES

The concept of an *equilbrium estuary* has arisen because of the commonly observed situation that the ratio of certain estuarine dimensions appears to be constant. It is considered there is a balance where the sedimentary infilling increases the currents in the estuary until they are strong enough to prevent further sedimentation. At the balance there would need to be as much sediment transported landwards on the flood tide as is carried seawards on the ebb. Therefore we might expect that flood dominated estuaries would tend to fill up with sediment brought in from the sea or coming down the river and becoming trapped. The shallowing water depth and the increased friction would then have to affect the tidal prism volume and the tidal response so that the currents diminish. Conversely, ebb dominated estuaries would tend to flush themselves clear of sediment, thus increasing the water depth, and this would have to alter the response towards flood dominance. For small tidal ranges virtually all estuaries appear to be ebb dominant, regardless of the extent of tidal flats. Strongly flood dominant estuaries are generally shallow, with high tidal ranges, and would tend to become shallower, whereas ebb dominated estuaries are relatively deep and would tend to become deeper. To produce an equilibrium estuary there would need to be a mechanism to take an estuary from flood dominance to ebb dominance and vice versa. For the former, sedimentary infilling would be required to increase the area of the tidal flats without affecting the mean depth of the estuary. This is a likely situation caused by the processes of tidal pumping that will be discussed later. Such infilling will have an affect on the tide range within the estuary as the cross-sectional area at the mouth will diminish as the tidal prism gets smaller, and this will affect the 'efficiency' of the mouth. The spring–neap variation in tidal range and volumes has also been considered by Friedrichs and Aubrey (1988), and they show that it is possible that the direction of dominance could change as a consequence. This situation has been demonstrated for the Tay estuary where there is a seaward flux of suspended matter at neap tides and an inward flux at spring tides (Dobereiner and McManus, 1983). We can speculate that this state may be an indication that the estuary has reached an equilibrium situation. However, this analysis has not considered the effects of river-borne influx of sediment, nor the effects of variable river discharge.

The concept of equilibrium was the basis of the empirical relationship of O'Brien (1969), who related the cross-sectional area A of the mouth of an estuary below mean sea level to the spring tidal prism P. He found that $A = cP^n$, where c is a constant, and $n \approx 1$. As P is actually a volume per tide, the constant c has the dimensions of an inverse velocity. The value that he derived for the constant $c = 2.10^{-5}\,\text{ft}^{-1}$ is equivalent to a velocity of $67\,\text{cm s}^{-1}$, the threshold of movement of sand. Thus the relationship specifies that an increasing tidal prism leads to an increase in the velocity at the mouth, which causes sand to move and the cross-section to increase until the velocity diminishes to the threshold value. Conversely, a decreasing prism will allow the mouth to infill until the tidal velocity through it increases sufficiently to just move the sand and maintain a constant area. Further work (e.g. Gao and Collins, 1994) has shown broad agreement, but with deviations that can be explained in terms of gravel or rock beds in the estuary mouth, or non-equilibrium conditions. The *O'Brien relationship* is one that is widely used as the basis of morphodynamic modelling, even to the extent of assuming such a balance locally at sections within an estuary.

A similar empirical relationship was developed by Van Dongeren (1992) for the tidal flat areas of the inlets of the Dutch Wadden Sea. He defined

$$A_{\text{flats}} = A_{\text{basin}}(1 - 0.025\sqrt{A_{\text{basin}}}) \qquad (3.3)$$

where A_{flats} is the area of the intertidal flats and A_{basin} is the total area of the estuary. There seems to be no physical basis for this relationship.

The depth, breadth and, consequently, the cross-sectional area of many estuaries appear to vary exponentially with distance down the estuary and this is thought to be an expression of equilibrium. For instance, the Ord River estuary in Australia, which is macrotidal with a range at the mouth of 5.9 m, has the dimensions (Wright *et al.*, 1973) $b_x = b_0 e^{-ax/L}$ and $h_x = h_0 e^{-bx/L}$ where $a = 4.28$ and $b = 2.76$.

Prandle and Rahman (1980) and Prandle (1991) have modelled the tidal response of estuaries varying in this way and find that there is a maximum amplification of the elevation when $a + 2b \approx 10$. Thus the Ord River fits this criterion. However, Wright *et al.* (1973) show that though the tidal amplitude decreases towards the head of the estuary and it is hyposynchronous, it is flood dominant. They explain the apparent ambiguity in terms of the sediment transport. The depth and width convergence tend to produce a distribution of frictional dissipation where the tidal wave does equal work per unit area of the channel bed, producing a state known as *entropy*. The work rate per unit area of the bed is proportional to the sediment transport rate and deposition or erosion of sediment requires a gradient in the sediment transport rate. Thus entropy implies that there are unlikely to be gradients in the rate of sediment transport along the channel and no resulting erosion or deposition within the estuary, even though sediment may be moving backwards and forwards during the tide. Consequently, though the flood dominance ensures sediment is moved landwards during the flood tide, an equal amount is carried seawards on the weaker, but longer duration, ebb tide. The Ord River, therefore, seems to be a well documented example of an equilibrium estuary. This is compared by Wright *et al.* (1973) with the neighbouring King estuary, which has a constant amplitude of tide from the mouth to a point only just downstream of the tidal limit. The banks are almost parallel and highly sinuous and the tide acts as a progressive wave. There is about equality between flood and ebb magnitudes. Thus it seems to be a synchronous estuary. Again, it is concluded that entropy is achieved within the estuary by the change along the estuary of the tidal prism, the depth and the width. Consequently, it appears that equilibrium can be achieved with a range of morphologies and tidal conditions, provided that entropy results. The entropy theory has not been widely considered and it is not clear which range of conditions satisfy it, but Wright *et al.* (1973) point out that with increasing river discharge relative to tidal prism the topographic convergence becomes less pronounced.

Chapter 4

Mixing

CHARACTERIZATION OF FLOW

A flow that does not vary in velocity with time is *steady* and one that does not vary in space is *uniform*. Measurements of velocity taken at a point are known as *Eulerian* velocities and those calculated by following a particle are *Lagrangian* velocities. In steady uniform flow they are the same, but in unsteady and non-uniform flows they are different. Estuarine flows are rarely both steady or uniform. However, many of the principles derived from laboratory flows which are steady or uniform can be applied.

Certain characteristics of the flow of homogeneous and stratified fluids in pipes and channels can be represented by two dimensionless numbers, the *Reynolds number* and the *Richardson number*. The Reynolds number compares the relative importance of inertial and viscous forces in determining the resistance to flow

$$Re = \frac{uD}{v} \tag{4.1}$$

where u is a velocity, D is a depth and v is the kinematic viscosity, the ratio of molecular viscosity to density (μ/ρ). In an unstratified fluid D will be the total depth of water and u the mean velocity. In this context, whether the flow is laminar or turbulent is determined by the value of the Reynolds number. Below $Re \approx 2000$ the flow can be laminar and above about 10^5 the flow is likely to be fully turbulent. Between these two points the flow is transitional and its character and the point at which it becomes fully turbulent depend largely on the roughness of the walls of the pipe or channel. Sternberg (1968) examined the flow in a number of tidal channels and found that fully turbulent flow occurred at an Re greater than 1.5 to 3.6×10^5 and that the flow over geometrically simple beds became fully turbulent at a lower Re than over beds

of complex roughness. In unstratified conditions in rivers and estuaries the flow
is always transitional or fully turbulent.

The competition between stratification and mixing play a crucial part in
estuarine dynamics because when the fluid is stratified the density gradient
resists the exchange of momentum by the turbulence and an extra velocity
shear is necessary to cause mixing. The *gradient Richardson number Ri* is a
comparison of the stabilizing forces of the density stratification to the
destabilizing influences of velocity shear and can be defined by

$$Ri = -\frac{g}{\rho}\frac{\partial \rho}{\partial z} \Big/ \left(\frac{\partial u}{\partial z}\right)^2 \tag{4.2}$$

For $Ri > 0$ the stratification is stable, for $Ri = 0$ it is neutral and the fluid
unstratified between the two depths, and for $Ri < 0$ it is unstable. When the
stratification exceeds a certain value turbulence will be damped, mixing will be
limited and the flow will be essentially laminar, even though it may be turbulent
in the homogeneous layers above and below. There have been many laboratory
and theoretical investigations of the mechanisms of formation and growth of
instabilities within the stratified interface, and the transition from laminar to
turbulent flow under conditions of uniform flow is generally taken to occur at
$Ri = 0.25$.

Because of the difficulties of measuring the gradients precisely, it is also
possible to define a *layer Richardson number*

$$Ri_L = \frac{(\Delta\rho/\rho)gD}{u^2} \tag{4.3}$$

where D is the depth of the upper layer flowing with a velocity u relative to the
lower layer and $\Delta\rho$ is the density difference between the layers. In this form the
Richardson number is a bulk number reflecting the characteristics of the whole
flow rather than the more detailed, localized gradient Richardson number
defined by Equation (4.2). The square root of the inverse layer Richardson
number is the *interfacial Froude number Fi*. The Froude number is an
alternative way of considering the influence of density stratification and *Fi* can
also be thought of as comparing the velocity of the flow with the velocity of
propagation of a progressive wave along a density interface. For a thin surface
layer of thickness h_1 and density ρ_1 flowing over a deep stationary layer of
density ρ_2, the propagation speed of a wave on the interface is $c =
(\rho_1 - \rho_2/\rho_2 \cdot g \cdot h_1)^{1/2} = (\Delta\rho/\rho \cdot gh_1)^{1/2}$. The *interfacial Froude number* is de-
fined as $Fi^2 = u^2/c^2$. Thus, slightly differently from Equation (2.5):

$$Fi = \frac{u}{\sqrt{(\Delta\rho/\rho)gh_1}} \tag{4.4}$$

In subcritical conditions, as u approaches c, a wave or perturbation can only
travel upstream very slowly and the wave energy accumulates, the wave

amplitude grows and at critical conditions when $Fi = 1$, the wave will break with energetic mixing. The thickness of the flowing layer will abruptly increase, with a decrease in velocity, and the flow goes from a supercritical to a subcritical Froude number. This situation is known as an *internal hydraulic jump*.

If both layers are relatively thin and flowing, the speed of the interfacial wave is

$$c = \sqrt{\left(\frac{\Delta\rho}{\rho}\right) \cdot \left(\frac{h_1 h_2}{(h_1 + h_2)}\right)}$$

The maximum velocity for subcritical flow is thus achieved when $h_1 = h_2$. This effectively sets a limit on the maximum two-way transport that can occur. At this stage the estuary is said to be *overmixed* (Stommel and Farmer, 1953). For this case

$$G^2 = \frac{u^2}{c^2} = u^2 \frac{\Delta\rho}{\rho} \cdot g \cdot \frac{h_1 h_2}{(h_1 + h_2)} = F_1^2 + F_2^2 \qquad (4.5)$$

Thus the combined or composite Froude number for such a two-way flow is the sum of the Froude numbers for the individual layers (Armi and Farmer, 1986; Farmer and Armi, 1986). When the lower layer is deep relative to the upper one, Equation (4.5) reduces to F_1^2. The composite Froude number is particularly useful in considering the dynamics over the shallow sill at the mouths of fjords.

When the density profile is not sharply divided into two distinct layers, a number of different internal waves speeds are possible, relating to different parts of the profile. Consequently it is possible that the flow may be subcritical for some parts while being critical for others. The internal wave-induced velocities in different parts of the water column can oscillate either in or out of phase. For a two-layer situation, an interfacial wave causes an increase in the surface layer velocity at the same time as a decrease in the lower layer velocity and vice versa. For three layers an increased thickness and a reduced velocity of the intermediate layer would cause thinning of the surface and lowest layers and increased velocities. More complicated oscillations and shears are possible. For a continuous stratification the critical Fi is 0.33.

An *estuarine Froude number* can be defined which relates the volume flow of freshwater R to the width b as well as the depth H of the control cross-section at the estuary mouth

$$F_e = \frac{R}{\sqrt{(\Delta\rho/\rho)gb^2h^3}} \qquad (4.6)$$

Thus a critical Froude number can be achieved by a constriction at the mouth as well as by a shallow sill.

MIXING IN STRATIFIED FLOWS

The mixing of salt and fresh water in estuaries is carried out by a combination of turbulence generated by shear at the sea bed and turbulence generated by shear at the halocline. These two effects vary in their magnitudes and timing during the tide, as well as in different estuarine types, as the stratification and the tidal velocities change. Let us first consider the mixing internally at the halocline.

Internal Mixing

The form of the instabilities created at the interface depends on the relative thicknesses of the velocity interface and the density interface, and the relative position of the centres of the gradient zones. For small current shear the density interface is smooth, but as the current in the upper layer, and consequently the shear across the interface, is increased, the interface becomes disturbed by waves with different mixing characteristics. The major distinction between them is in terms of Richardson number. The processes have been described by Thorpe (1973), McEwan (1983), Yoshida (1986) and reviewed by Fernando (1991).

Entrainment

At $Ri > 1/4$ the instabilities take the form of 'cusps', or progressive interfacial *Holmboe waves*, which grow in height and become sharper crested. Eventually they break and wisp-like elements of the denser water are ejected from the crests into the lighter layer above (Figure 4.1). This process of breaking interfacial waves causes *entrainment* and the critical velocity above which this process occurs must be related to the degree of density stratification. Little passage of salt into the upper layer occurs until the waves break and the amount of entrainment will decrease as Ri increases.

Though the breaking of each wave is a discrete process, it is so continuous in space and time that, when averaged, the loss of water into the upper layer can be likened to vertical flow, an *entrainment velocity* u_e. The ratio of u_e to the

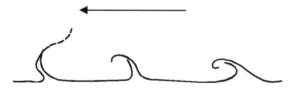

Figure 4.1 Diagrammatic representation of interfacial Holmboe waves on a halocline with entrainment from the denser layer.

velocity in the upper layer u varies inversely with Ri, i.e. $u_e/u \approx Ri^{-1}$. Thus the rate of entrainment will increase with increasing velocity difference between the layers, i.e. with decreasing Ri.

At high Ri the impact with the interface of turbulence in the overlying or underlying flow is limited. Eddies flatten, but distort the interface and can generate small internal waves that can then grow when acted upon by the shear.

Because of problems of scaling and the unsteady nature of the flow, results from laboratory experiments involving entrainment cannot be applied with certainty to natural estuaries. However, similar processes are probably operating naturally. A vivid description of mixing in a natural estuary has been given by Farrell (1970) describing the Saco River in Maine: 'Ninety minutes after high water, a strong fresh-water current, with a surface flow of up to 5 feet per second, had developed, while less than 6 feet down current velocities at five stations were less than 1.0 feet per second. The effect was most pronounced at Station 9 where the velocity of 5 feet per second existed only in the top 3 feet of the water column. So pronounced was this fresh-water–salt-water boundary that a diver approaching it below in the virtually still seawater could observe the brown river water flowing overhead in violent turbulence. The actual boundary was so well defined that it was limited to a 6-inch width. Occasionally, vortices distorted the boundary to a depth of 4 or 5 feet below the boundary surface. This caused waveform undulations to spread upstream until the vortex subsided and the straighter plane of shearing flow re-established itself'.

Thus entrainment is a one-way process in which a less turbulent water mass becomes drawn into a more turbulent layer. Because of the vertical movement of salt in the entrained water, the potential energy of the water column is increased, the energy for this coming from the kinetic energy of the horizontal flow. As a consequence of entrainment of extra volume into the upper more turbulent layer, the discharge of the more turbulent layer increases downstream.

Turbulent Diffusion

At $Ri < 1/4$ the waves have a different form, and they are called *Kelvin–Helmholtz instabilities* or K–H waves. In estuaries these waves would have wavelengths of the order of a metre or so. As a consequence of the velocity shear they rapidly steepen, overturn and tend to roll up into a spiral or 'cat's-eye' structure (Figure 4.2). The centre of the vortex is at the level of the maximum gradient and it breaks down into a turbulent core called a *'billow'*, into which water from both above and below the interface is mixed. The billow becomes drawn out in the direction of the shear into an obliquely inclined patch separated above and below by stable gradients. Thus a stepped profile (microstructure) develops. The overall density interface becomes thicker than

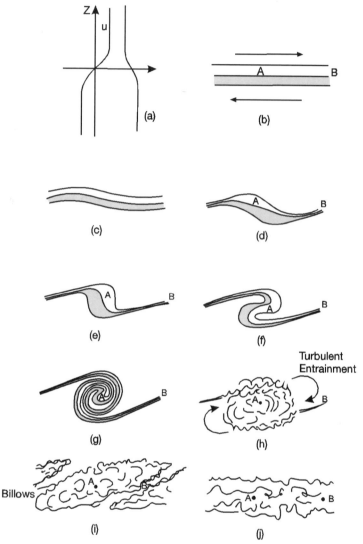

Figure 4.2 Growth of disturbances on a sheared interface. (a) Velocity and density profiles. (b) Indicative fluid layering. (c–j) Growth, roll-up and breakdown of an instability. Reproduced by permission of the American Geophysical Union from Thorpe (1969).

the velocity interface and turbulence in the bounding flows can affect the mixing. The turbulent eddies in both the overlying or underlying flows impact with the interface causing it to deflect, and its rebound causes 'splashing'. This is most effective when Ri is low, and acts to move water and salt upwards from the interfacial layer into the overlying flow, as well as downwards into the flow below. Consequently, the net mixing is the sum of the two. The mixing resulting from this turbulence is therefore a two-way process in which equal volumes of water are exchanged between the two layers. This is known as *turbulent diffusion* and requires the presence of turbulence in both layers. Though there is no net exchange of water, salt is transported upwards and the potential energy of the water column is again increased.

Thus the essential difference between entrainment and diffusion is the degree of turbulence in the two layers. If the turbulence is the same in both upper and lower layers there is minimal entrainment, all the mixing being by turbulent diffusion. If the lower layer is static, then there is no turbulent diffusion across the interface and the mixing is entirely by entrainment. Because of the unsteady nature of the flow, short bursts of intense turbulent mixing are separated by periods where entrainment predominates. When averaged over space and time, both methods of mixing work together, their relative importances depending on the degree of turbulence in the lower, saline layer, as well as the *Ri*. As estuaries are mainly stratified and turbulent, mixing will be carried out by both entrainment and diffusion. To a first approximation the ratio of their contributions will vary with the ratio of river discharge to tidal prism volume. Entrainment will be active at high *Ri*, and turbulent diffusion at low *Ri*. Consequently, highly stratified estuaries are dominated by entrainment, while in partially stratified estuaries turbulent diffusion will be dominant.

Internal Waves

An additional factor that is important in tidal estuaries is the presence of *internal waves*: wave features on the interface with wavelengths of tens of metres. In a gradually accelerating flow of stratified water over irregularities in the sea bed, a series of *lee waves* will form on the interface as the stratification tries to conform to the sea bed shape. The upstream phase velocity of the waves just equals the flow velocity. Thus they are static in steady flow. With increasing velocity the waves with shorter wavelengths are gradually displaced downstream by waves of larger wavelengths and amplitudes (larger phase velocities). As critical internal Froude number conditions are approached, the wave energy is concentrated near to the depth change, the amplitude of the wave grows and it breaks as an internal hydraulic jump, creating active mixing that is then advected downstream. If critical conditions are not reached the phase velocity of the waves causes them to propagate upstream as the mean flow diminishes. The internal waves can subsequently break if they encounter critical conditions further upstream, particularly if the stratification or layer thickness decrease. These waves will interact with the shear and locally the Richardson number can become critical and billows will appear, first at the crests and troughs of the internal wave. Geyer and Smith (1987), with detailed field measurements, have shown that internal wave interaction with the shear will raise the critical Richardson number to 0.33, so that turbulent mixing can occur even when the time–mean gradients remain above the critical value. They also found that the turbulent mixing resulted in an increase in the vertical extent of the gradient zone, with the magnitudes of the gradients staying constant and producing an equilibrium pycnocline thickness

$$\delta = \frac{0.33(\Delta u)}{g(\Delta \rho / \rho)}$$

$\Delta \rho$, Δu are the differences in density and velocity across the interface. As internal waves are intermittent features, the critical Richardson number is likely to be between 0.25 and 0.33.

In principle, internal waves can be formed on both ebb and flood tides, but in practice they are more likely on the ebb because of the increasing water depths downstream. Thus mixing can be localized around areas of rough topography or sudden depth changes.

With modern high frequency echosounders it is possible to visualize the mixing in a direct way. Examples of this are shown in Dyer (1988a), Dyer and New (1986), Geyer and Smith (1987) and Sturley and Dyer (1992). Figure 4.3 shows several long internal waves being formed at a deepening of the estuary. Superimposed on them are Kelvin–Helmholtz waves that are concentrated on their upstream faces. These waves, which have typical wavelengths of a few metres, create a mixed zone on the halocline and inclined structures sloping upwards in a downstream direction. A vertical salinity profile through the structures often penetrates several layers, each of which is homogeneous and separated by zones of higher gradient. Sometimes the internal waves can be seen at the surface as a series of patches of smooth and rippled water elongated transversely across the estuary, and they can be seen to propagate along the estuary. The rippled water coincides with the crests of the internal waves.

Boundary Layer Turbulence

The frictional drag at the sea bed produces a velocity shear, the near bed flow being slowed down considerably relative to that higher up, and turbulence is produced by the flow over and around the roughness elements of the bed. Thus the energy for the turbulent mixing is derived from the kinetic energy dissipated by the water flowing across a rigid rough boundary. The velocity profile close to the bed for unstratified conditions conforms to the formula

$$\frac{u}{u_*} = \frac{1}{\kappa} \ln \frac{z}{z_0} \qquad (4.7)$$

This is known as the *von Karman–Prandtl equation*. u_* is the friction velocity $(= (\tau_0/\rho)^{1/2})$, z_0 is the roughness length, which relates to the hydrodynamic roughness of the sea bed, and κ the von Karman constant $= 0.4$. This equation has been shown to be valid for homogeneous flows over flat beds and is

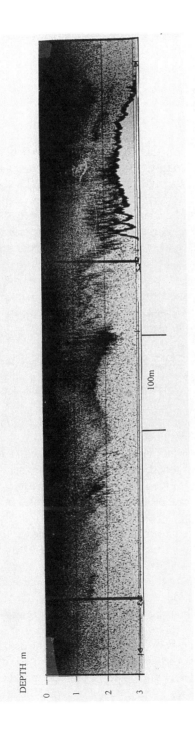

Figure 4.3 Echosounder trace taken in the Tees estuary, showing Kelvin–Helmholtz waves produced by shear of the surface layer flowing from right to left over a denser lower layer. The sea bed with sand waves appears as the dark feature at the lower right. Reproduced by permission of Humana Press from Dyer (1989).

commonly used to determine the bed shear stress τ_0 from the velocity profiles. The roughness length has been empirically related to the sea bed composition and form. The form of the velocity profile deviates from Equation (4.7) for unsteady flows, in density stratification and over uneven beds (see Dyer, 1986). For stratified flows the profile is modified to

$$\frac{u}{u_*} = \frac{1}{\kappa} \left(\ln \frac{z}{z_0} + \frac{az}{L} \right) \tag{4.8}$$

This shows that in stratification the velocity at any particular height will be faster than in homogeneous conditions. The value of $a \approx 5$, and the value of z/L depends on the *Ri*. *L* is the Monin–Obukov length, a length scale for mixing. Anwar (1983) found that Equation (4.8) represented the velocity profiles during an accelerating current within 2 m of the bed when *Ri* varied between about 0.025 and 0.3. The value of a/L was reasonably constant at 0.015–0.025 and a averaged 11.4. During well-mixed conditions the profiles conformed to Equation (4.7). The corresponding salinity profiles were self-similar when normalized using the near bed and near surface salinities.

Alternatively, experiments have shown to a good approximation that the bed shear stress is proportional to the square of the velocity in the boundary layer. This gives the *quadratic friction law*

$$\tau_0 = \rho C_d u^2 \tag{4.9}$$

The drag coefficient C_d can be related to the bed roughness and to the value of z_0 (see Dyer, 1986).

Measurement of the velocity vectors in a turbulent fluid show random fluctuations which can be distinguished from the steady advective flow which moves the turbulence along. Flow visualization studies close to the boundary have shown that turbulence is associated with transverse vortices rotating with the flow (Nychas *et al.*, 1973; Offen and Kline, 1975; Robinson, 1991). Very near the bed, for smooth bed conditions, the flow consists of low and higher velocity streaks. One of the streaks is lifted up in a three-dimensional disturbance and a local recirculation cell is formed beneath the streak, which develops into a vortex. As the vortices are advected downstream they develop, producing a vortex that has a 'horseshoe' or 'hairpin' shape, with the ends at the boundary and an inclination of 16–20° to the horizontal. They travel outwards away from the boundary as they are advected downstream, growing larger, though less intense. The process is illustrated in Figure 4.4. The vortices travel downstream at about 0.8 of the mean flow and they gradually break down because of being stretched by the velocity gradient near the bed and by viscous dissipation. As the vortices are being generated intermittently at various places on the bed, at any one position different parts of the horseshoe would be seen at different times at each height above the bed, resulting in the

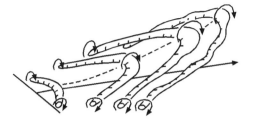

Figure 4.4 Diagrammatic representation of the growth of a horseshoe vortex in the sea bed boundary layer with progress downstream.

characteristically chaotic velocity pattern. In rivers and estuaries at times of high flow these vortices erupt at the surface as 'boils' (Jackson, 1976).

At any position near the bed, the downward flow at the front of the vortex would be seen as a high velocity 'sweep' directed towards the bed, with a higher than average horizontal downstream component for a short while. The trailing edge of the vortex would be seen as a short period of lower velocity with a component directed away from the bed, an 'ejection'. The passage of the vortex produces a turbulent burst containing a sweep/ejection cycle.

However, turbulence needs to be considered in a stationary frame to quantify the exchanges of mass and momentum. The velocities can be separated into three orthogonal components, u, v and w, in the streamwise horizontal, lateral and vertical directions, respectively. Consider in Figure 4.5 the situation without a tide, so that $u = 0$. The velocity can be separated into a time mean flow $\bar{u}$ and a turbulent deviation u'. Thus the observed instantaneous velocity $u = \bar{u} + u'$. Similarly, $v = \bar{v} + v'$ and $w = \bar{w} + w'$. Fluid particles

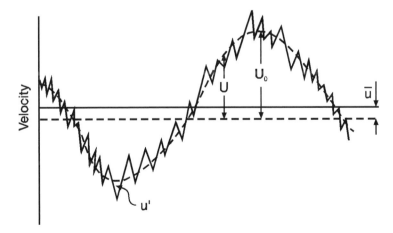

Figure 4.5 Definition sketch for the mean, tidal and turbulent velocity components from a velocity time series.

moving upwards carry a momentum deficit into the faster flow above. Conversely, fluid moving downwards carries excess momentum into the slower moving layers below. Even though the time average of both u' and w' are zero, by definition, and there can be no exchange of water, the average of their product $\overline{u'w'}$ is non-zero, and there is an exchange of momentum across the horizontal plane parallel to the flow direction. This momentum exchange is the shear stress τ acting on the plane. The momentum flux $F = \overline{\rho u w} = \rho \overline{u}\,\overline{w} + \rho \overline{u'w'}$. The first term is the advective flux, the second the turbulent momentum flux, i.e. the shear stress. The other terms $\overline{u'\overline{w}}$ and $\overline{\overline{u}w'}$ are by definition zero. Thus $\tau = -\rho \overline{u'w'}$. There are similarly shear stresses on the two vertical planes due to interactions represented by $-\rho \overline{v'u'}$ and $-\rho \overline{v'w'}$. The turbulent shear stresses are called the *Reynolds stresses*.

The Reynolds stresses can be thought of as composed of four contributions, an ejection event ($u' < 0$, $w' > 0$), a sweep event ($u' > 0$, $w' < 0$) and outward and inward interaction events ($u' > 0$, $w' > 0$ and $u' < 0$, $w' < 0$). Only the ejection and sweep events provide a positive contribution to the Reynolds stress. It has been estimated that as much as 70% of the stress may be caused by the ejections, even though they exist for only 20% of the time. Heathershaw (1974) reports field measurements that show that 57% of the stress is developed in only 7% of the time.

A similar analysis can be carried out for the vertical salt exchanges. Using $s = \overline{s} + s'$, the vertical turbulent salt flux will be the time mean of the product of the salinity and the vertical velocity, $\overline{s'w'}$. This is known as the *turbulent* or *Reynolds flux* of salt. A positive contribution to the vertical salt flux only occurs when $+ w'$ is correlated with $+ s'$, or $- w'$ to $- s'$. West and Shiono (1985, 1988) found that with increasing Ri, the vertical flux decreased relative to the horizontal, and there was evidence that u' and w' tended to become $\pi/2$ out of phase, and u' and s' π out of phase. This indicates that with increasing Ri wave-like motions replace the random turbulence which is inhibited by buoyancy effects, thus fitting the descriptive models above.

The intensity of the turbulence is measured by the ratio of the root mean squared velocity to the mean value. Near the sea bed the turbulence intensity of the horizontal flows is generally about 8–10%, with the vertical intensities being about 40% smaller. When stratification is present the vertical intensity diminishes relative to the horizontal, but when internal waves are present the vertical fluctuations are relatively enhanced, but do not contribute much to the vertical exchanges as they are 90° out of phase with the horizontal fluctuations.

Estuarine Mixing

In most estuaries the mixing is a combination of internally generated and boundary generated turbulence, which varies in relative magnitudes in space and time. The result of this can often be seen in the form of the salinity profiles in highly stratified and in partially mixed estuaries (Figure 4.6). The overall

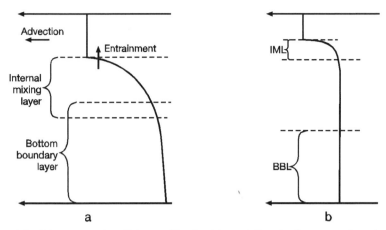

Figure 4.6 Diagrammatic salinity profiles for (a) partially mixed estuary where internal and bottom mixing interact and (b) highly stratified estuary where internal and bottom mixing are separate.

effectiveness of the turbulent mixing will be greatest when the velocity shear in the turbulent sea bed boundary layer overlaps with the mixing occurring at the interface. This will cause a gentle gradient at the bottom of the halocline and the maximum salinity gradient to be at the upper part of the interface. Where the two do not overlap, the maximum gradient will be near the middle of the interface and the bottom layer will be homogeneous. Abraham (1980) has shown that in well-mixed estuaries the turbulence is primarily boundary generated, with internally generated turbulence only important at about slack water. In highly stratified estuaries the turbulence is generated at the interface. For the Rotterdam Waterway the ratio of the internally generated turbulence to the boundary generated turbulence was $> 1/2$ during the ebb tide and $> 1/5$ for the flood tide. For the Tees estuary, Lewis (1996) concludes that the generation of turbulent energy at the bed is 5–30 times that at the interface.

The form of the spectra of the salinity variations at about mid-depth in the water can be used to distinguish between the effects of inactive wave-like motions and turbulent fluctuations. Dyer (1983) proposed practical limits for mixing regimes based on the spectral shapes and the layer Richardson number [Equation (4.3)] using the full water depth as the length scale. This, therefore, implicitly includes both internally generated and boundary mixing. It was proposed that for $Ri_L > 20$, bottom generated turbulence was ineffective in decreasing the stratification; for $20 > Ri_L > 2$, mixing is increasingly active and for $Ri_L < 2$, fully developed mixing occurs. Dyer and New (1986) used this approach to illustrate that mixing occurred preferentially in the shallow parts of a cross-section of Southampton Water on the ebb tide, and increased towards spring tides.

Nevertheless mixing is inefficient. A maximum of about one-quarter of the internal wave energy can be converted into increased potential energy by wave breaking and in general 2–5% of the barotropic energy of the tide goes into increasing the potential energy of the water column by mixing.

Mixing Coefficients

Turbulence is difficult to measure and so a means of parameterizing it in more easily measured terms is sought. It is common to treat the turbulent exchanges as a gradient process rather like molecular diffusion or conduction and proportional to the mean gradients. Thus

$$-\rho(\overline{u'w'}) = \rho N_z \partial \overline{u}/\partial z \qquad (4.10a)$$

$$-\rho(\overline{w's'}) = \rho K_z \partial \overline{s}/\partial z \qquad (4.10b)$$

The coefficients are known as eddy coefficients, K_z being the *eddy diffusion coefficient*, or *eddy diffusivity*, and N_z the *eddy viscosity*. They have the dimensions of $m^2 s^{-1}$. These coefficients, however, are unlike the molecular coefficients in that they are not intrinsic properties of the fluid. They will vary in time and in space depending on the flow conditions, particularly *Ri*, the boundary roughness and the distance from the boundary. Determination or choice of the magnitudes to be used is an important question for numerical modelling.

In homogeneous water the exchange coefficients are considered to be equal, i.e. $K_z = N_z$. In the presence of a density gradient the turbulence has to do work against the gradient, thus turning kinetic energy into potential energy. Thus both K_z and N_z are reduced below their homogeneous value. However, N_z is reduced less than K_z, because velocity shear occurs below the interface and there is a small momentum transfer when internal waves are present even when there is little mass transfer. The fraction of turbulent energy that is taken up in vertical mixing, i.e. the efficiency, is represented by the *flux Richardson number*

$$R_f = K_z/N_z \cdot Ri$$

The ratio K_z/N_z is the *Prandtl Number*. The flux Richardson number reaches a maximum value of about 0.2 at the critical gradient Richardson number of 0.25 (Linden, 1979; McEwan, 1983). A consequence of having a maximum in the efficiency is that mixing will tend to generate an interface that will have an *Ri* of about 0.25, and further mixing will increase the interface thickness rather than decrease the *Ri*. This has been explored by Posmentier (1977) as the mechanism for generating step-like density profiles.

Various researchers have tried to determine relationships between the eddy coefficients and the *Ri*. Field determinations of relationships have shown a large scatter of results. These have been summarized by Dyer (1988a). The form of the curves is shown in Figure 4.7. The most commonly used relationships are those of Munk and Anderson (1948), which are

$$K_z = K_0(1 + 3.33\,Ri)^{-3/2} \qquad\qquad (4.11a)$$

$$N_z = N_0(1 + 10Ri)^{-1/2} \qquad\qquad (4.11b)$$

K_0 and N_0 are the value of the eddy coefficients at zero stratification. Though there is considerable variation, the various curves have the common feature of a point of inflection at an Ri of about 0.25.

To use the above relationships, values for N_0 are needed (or K_0, as they are equal). In the boundary layer close to the sea bed where the velocity profile is logarithmic and the shear stress constant, $N_0 = \kappa u_* z$. If the logarithmic profile extends to the surface and the shear stress distribution is linear, becoming zero at the surface for no wind, the profile of N_0 is parabolic with a maximum at mid-depth of $N_0 = \kappa u_* h$, where h is the water depth. Bowden *et al.* (1959) have shown that at mid-depth in an unstratified flow with a tidal current of amplitude U_0, $N_0 = 2.5 \cdot 10^{-3} U_0 h$. The value of the constant is likely to vary for different bottom conditions. At the sea surface the wind stress through the action of waves will play an important part in determining N_0. Kullenberg (1977) found that $N_0 = 4.1 \cdot 10^{-5} W^2$, where W is the wind speed in $\mathrm{m\,s^{-1}}$ at 10 m height.

The diffusion coefficients can be calculated directly from the diffusion of dye patches over short distances in the flow. With a continuous injection of dye a plume is formed that spreads downstream because of the turbulent mixing (Figure 4.8) and the variance of the distribution of dye concentration, σ^2, increases as time to the power 2 to 3. When the diffusion coefficient is independent of time or of distance, then

$$\sigma_x^2 = 2K_x t_x. \quad \text{Similarly,} \quad \sigma_z^2 = 2K_z t_z, \quad \sigma_y^2 = 2K_y t_y$$

If σ_z is taken as 5 m, then for $K_z = 1\,\mathrm{cm^2\,s^{-1}}$, t_z would be 35 hours; for $K_z = 10\,\mathrm{cm^2\,s^{-1}}$, t_z would be 3.5 hours; and for $K_z = 100\,\mathrm{cm^2\,s^{-1}}$, t_z would be 21 minutes. Therefore when K_z is about $100\,\mathrm{cm^2\,s^{-1}}$, complete mixing in 10 m water depth would take about half an hour. For a weakly stratified flow the vertical mixing time-scale would be of the order of a tide, and for strong stratification several days.

Dispersion

One feature of turbulence is that it occurs at the same time as shear in the mean currents. As turbulence cannot be measured at many points simultaneously, a spatial average has to be taken. This assumes that the turbulence is representative of the larger volume, and the advection on the mean current is also averaged over the same volume. Any variations in the mean currents

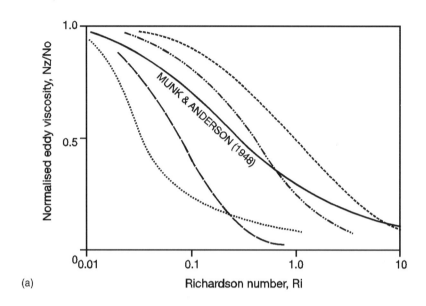

(a)

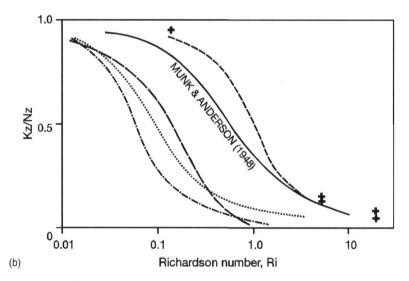

(b)

Figure 4.7 Relationship of eddy coefficients with gradient Richardson number. (a) Various theoretical and experimental curves of normalized eddy viscosity against *Ri*. (b) Various curves of Prandtl number versus *Ri*. Reprinted with permission from Dyer, K.R. (1988a). Tidally generated estuarine mixing processes. In: *Hydrodynamics of Estuaries*, Vol. 1 (Ed. B.J. Kjerfve). Copyright Lewis Publishers, an imprint of CRC Press, Boca Raton, Florida.

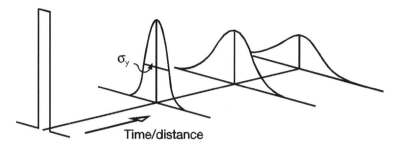

Figure 4.8 Representation of the lateral diffusion of an input of marked fluid with time/distance downstream. Height of the curves is proportional to concentration.

across the volume then appear to be part of the turbulence. In modelling terminology this would be called *sub-grid scale turbulence*, with the volume over which the average is taken being the grid scale. The simplest way of describing the effect is to consider a shallow homogeneous flow in which there is a velocity gradient (Figure 4.9). If a vertical streak of dye is put in instantaneously, it will distort through time, with the near surface dye travelling faster than that below. If there is vertical mixing due to turbulent diffusion, the surface dye at the leading edge will be mixed downwards and the near bed dye at the trailing edge will be mixed upwards. The overall effect is that the dye streak will spread out as it moves downstream. Considered in terms of the depth mean velocity, the spreading is called *dispersion*. Dispersion is considered as a gradient process, as is diffusion, and its magnitude will depend on the shear in the horizontal velocity, combined with the vertical turbulent diffusion. The dispersion increases as the rate of diffusion decreases, and K_x is proportional to K_z^{-1}, where K_x is a dispersion coefficient and K_z a

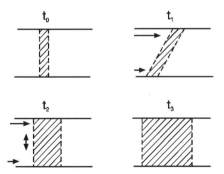

Figure 4.9 Dispersion of marked fluid in a shear flow. t_0, Initial distribution in shear flow without vertical mixing; t_1, Distribution after a time interval. Between t_1 and t_2 vertical mixing is switched on. t_2, Distribution with shear and vertical mixing; t_3, after a further time interval with shear and mixing.

diffusion coefficient. Conceptually, dispersion is reversible if the shear is reversed, whereas diffusion is not. We must always be clear about the distinction between them, because the word diffusion is often used in situations where the processes are dispersive. Engineers generally use the symbol D_x for the dispersion coefficient.

Dispersion coefficients have been calculated or measured for a variety of situations. The dispersion of dye in unidirectional flow in pipes has been studied by Taylor (1954), who showed that in a pipe of radius a

$$K_x = 10.1au_*$$

where u_* is the friction velocity. For open trapezoidal channels Elder (1959) found that

$$K_x = 5.93hu_*$$

where h is the depth of flow.

Experimental results for natural streams, however, show the observed K_x is an order of magnitude higher than these theoretical estimates. The theoretical estimates are for unidirectional flow, but Bowden (1965) has shown that, with certain assumptions, the dispersion coefficient for alternating flow is about half that for unidirectional flow. For a tidal current of amplitude U_0, $K_x = 0.15U_0h$ (Bowden, 1963). In rivers the K_x has been shown to be as great as $K_x = 1000hu_*$ (Fischer, 1973, 1976). The magnitude of the coefficient depends on the presence of bends which produce secondary currents and affect the cross-sectional mixing time-scales.

Averaging

Turbulence Averaging

The presence of tidal movements and of turbulence introduce problems that can only be resolved by averaging. We need to be able to separate the influences of river flow, tidal oscillation and turbulent fluctuations to understand the interactions and processes in an estuary.

An instantaneous measurement of velocity u can be represented by three components

$$u = \bar{u} + U + u' \tag{4.12}$$

where $\bar{u}$ is the mean velocity over a tidal cycle, U is the tidal variation and u' is the short period turbulent contribution (Figure 4.5). In many estuaries U can be represented moderately realistically by a trigonometrical function such as $U = U_0\sin\omega t$. It is difficult for measuring instruments to record the instantaneous velocity over long periods and at once averaging has to be introduced.

Then the observed velocity $u = \bar{u} + U$

$$\text{i.e.} \quad u = \bar{u} + U_0 \sin\omega t \qquad (4.13)$$

Taking values over a tidal cycle from $\omega t = 0$ to 2π then

$$\frac{1}{2\pi} \int_0^{2\pi} u = \bar{u} + \frac{1}{2\pi} \int_0^{2\pi} U_0 \sin\omega t \, \mathrm{d}t = \bar{u}$$

A similar analysis can be carried out for salinity measurements, but with the exception that S is generally about 90° out of phase with velocity and would be represented by $S = S_0 \cos\omega t$.

To satisfy Equation (4.13) a suitable length of time for each observation must be chosen to separate the essentially harmonic tidal frequencies and the random turbulent frequencies. To do this the two components must occupy different frequency ranges in the energy spectrum, there must be a gap between the two, with the averages taken over a length of time equivalent to the middle of the gap.

To remove the most significant of the turbulent fluctuations we need to measure over a period of about 100 s. This agrees with a rough indication which can be achieved by saying that we have to average over a time equivalent to the passage of water a distance, equal to about ten times the depth, down the estuary past a point. If the depth were 10 m and the current 100 cm s^{-1}, the averaging time would need to be 100 s. It also appears that we really need to average over a longer time when the tidal current is low than when it is fast and also we need to measure over a longer period with increasing height above the bottom.

Experience has shown that an averaging time of about 1–1.5 minutes is generally reasonable. In most cases this averaging is mainly visual and may be open to considerable errors. Once this averaging is complete, then the effect of longer period turbulence appears in the tidal terms.

Tidal Averaging

To obtain the value for the mean flow we need to remove the oscillation of the tide. Considering Equation (4.12), the average of U over a tide will be zero. We also need to take account of the variation of depth during the tide, otherwise there will be certain heights above the sea bed where there will only be a velocity reading at high water. The process of averaging to preserve correct discharge needs to be carried out in a careful sequence. This has been detailed by Kjerfve (1979), and is outlined below, assuming that the turbulence has been averaged out in the measurements. If this has not been done adequately, the errors are transferred into the tidal values.

1. To take account of the depth variation we normalize the measured profiles so that a continuous profile can be drawn in terms of z/h, so that the surface has a value $z/h = 0$, and the bed $z/h = 1$. Values at each measurement time need to be interpolated for intervals of z/h. If these intervals are $z/h = 0.05$, 0.2, 0.4, 0.6, 0.8, 0.95, the points are equally disposed about the average depth $z/h = 0.5$. However, to obtain a depth average, the values at z/h of 0.05 and 0.95 need to be half-weighted as they represent the flow in a fraction of only 0.1 of the water depth rather than 0.2. Thus the depth average would be

$$\langle U \rangle = 1/5\{(1/2\,u_{0.5}) + u_{0.2} + u_{0.4} + u_{0.6} + u_{0.8} + (1/2\,u_{0.95})\} \qquad (4.14)$$

2. To do the averaging with good confidence limits we need a large number of equally spaced observations over one or two complete tidal cycles. For adequate representation of the mean, observations every half-hour can suffice, though in some cases it is possible to get away with hourly readings. As the semi-diurnal tidal cycle is 12 h 25 min long, using solar hours will not give the required constant interval. Consequently we need to use lunar hours, each of which will be 2.1 min longer than the solar hour. Values of the velocity are therefore interpolated at each of the normalized depths for each lunar hour, or half-hour.
3. A matrix can then be constructed with columns of velocity values for each lunar hour (half-hour), and rows of velocity values at each normalized depth increment. Using Equation (4.14), the average of each column will give the depth mean velocity at each lunar interval $\langle U \rangle$, and the average of each row the tidal mean velocity at that normalized depth increment $\bar{u}_z$. Averaging the former through time, or the latter through the depth, will give the depth and tidal mean velocity $\langle \bar{u} \rangle$.
4. The same process can be carried out for salinity, suspended sediment concentration or any other constituent.

There will be errors in the tidal mean which relates to the sampling intervals. These are likely to decrease as the time interval is reduced. A 30 min sampling interval yields flux estimate errors of 30% relative to 5 min samples (Reed, 1987). It is normally difficult to maintain high frequency of sampling over a full tide, or more, so compromises have to be accepted.

Mean Flow

The mean flow $\bar{u}$ is the *Eulerian mean* flow, or the *non-tidal drift*. Its value, obtained in the above Eulerian way and averaged over the cross-section, need not be equivalent to the river discharge. Because of the tidal fluctuation in cross-sectional area, the volume transport for a unit mean velocity will vary throughout the tidal cycle. If $u = \bar{u} + U_0 \sin \omega t$, where u does not vary with width or depth, and $A = \overline{A} + A_0(\cos\omega t + \theta)$, where $\overline{A}$ is the mean cross-sectional area over a tidal period

$$\text{River discharge} = \frac{1}{T}\int_0^T Au\,dt$$

$$= \overline{A}\overline{u} + \frac{1}{2\pi}\int A_0\cos(\omega t + \theta) \cdot U_0\sin\omega t \cdot dt \qquad (4.15)$$

The first term on the right-hand side is the mean advection on the non-tidal drift and the second term is known as the *Stokes drift*. This arises because the discharge per unit time near to high water is greater than that near low water. If the tidal wave in the estuary is a standing wave and slack water occurs at both low and high water then $\theta = 0$, the Stokes drift is zero and the river flow equals the mean velocity $\overline{u}$. If the tidal wave is purely progressive then $\theta = \pi/2$ and the Stokes drift becomes $-\frac{1}{2}A_0U_0$ and the river discharge $R = \overline{A}\overline{u} - 1/2A_0U_0$. The velocity equivalent to the river discharge is that measured by following a floating or suspended particle, and is the *Lagrangian velocity*. Consequently the observed mean flow or non-tidal drift $\overline{u}$ can exceed the movement due to the river discharge divided by the cross-section area by an order of magnitude (Pritchard, 1958). The additional part of the non-tidal drift is an outwards compensation for inward mass transport caused by the partially progressive nature of the tidal wave. In the Columbia estuary slack water lags the maximum and minimum cross-sectional area by 1–1.5 hours even at low river stage. As a consequence 30% of the non-tidal drift is compensation for the inward mass transport (Hansen, 1965).

The calculated tidal mean flow need not be constant from one tide to the next. This may be the result of the inequality in the diurnal tide which means that the two semi-diurnal tides in the day have different ranges. These differences will also change between the spring to neap, and the neap to spring periods. Additionally, there will be large variations in the mean flows due to weather effects. These will be dealt with in detail in Chapter 9.

Chapter 5

Cross-channel Effects

So far we have concentrated on the variations along the estuary and assumed that the variations across the cross-section are small. However, the flows are not constant across the cross-section and this is likely to be the norm rather than the exception. The flow along the estuary is affected by bends and changes in water depth which tend to make the current concentrate on one side or the other and to spiral as the water travels downstream. This involves lateral and vertical components in the flow.

VERTICAL VELOCITIES

Though it is possible to measure instantaneous vertical velocities, it is at present difficult to obtain sufficient measurements over an area and in time to be able to calculate the mean vertical velocity. The situation has been improved by the development of the acoustic doppler current profiler (ADCP). These instruments use the Doppler shift of an acoustic signal backscattered from particles suspended in the water as a measure of their velocity. The acoustic beam is transmitted at an angle through the flow, either from a sea bed mounted transducer or one attached to a survey ship. Returns are recorded throughout the flow, but losses occur near the bottom and top part of the flow. After correction for the ship's motion the calculated velocities can be presented as three orthogonal components, one of which will be in the vertical plane. As the ship takes some time to travel across the estuary there has to be an assumption of spatial and time continuity in the velocity distribution. Examples of the use of this technique have been published by Jay and Musiak (1996) and Nepf and Geyer (1996).

An alternative method which can be used on tidally averaged data involves the continuity of volume. As velocity is a vector with a magnitude and direction, it can be represented by three components acting on mutually perpendicular axes. The x direction is taken longitudinally positive downstream, the y axis is lateral, positive to the right, and the z axis is vertically downwards. The values of longitudinal mean velocity $\bar{u}$ and lateral mean velocity $\bar{v}$ can be obtained from measurements at stations on cross-sections of the estuary. Then [similarly to the derivation of Equation (6.1)]

$$\frac{\partial \bar{u}}{\partial x} + \frac{\partial \bar{y}}{\partial y} + \frac{\partial \bar{w}}{\partial z} = 0 \qquad (5.1)$$

This is the equation of *volume continuity* and simply formalizes the fact that what goes in must come out. Any slowing down of water in the longitudinal sense requires additional discharge laterally or vertically. Using the horizontal mean velocities at stations on two cross-sections, then Equation (5.1) can be solved, provided one boundary condition is known, i.e. that $\bar{w}$ is known at one depth. If tidal rise and fall have been eliminated by averaging over a tidal cycle, then a valid boundary condition is that $\bar{w}$ is zero at the sea surface where z is zero. The distribution of $\bar{w}$ with depth can then be calculated by stepwise integration from the surface downwards.

The results of this sort of calculation are shown in Figure 5.1. Volume continuity was assumed between two cross-sections, on each of which three stations were occupied simultaneously for a tidal cycle at high and low river flow. The cross-section was on a left-hand bend looking downstream. At high flow the flow was downwards at the right-hand station and upwards at the left. In the middle the vertical velocities were small, but there was a strong lateral flow towards the right at the surface and towards the left near the bed. The result is a circulation in the cross-sectional plane, a phenomenon known as *secondary flow*, in this case in a clockwise sense.

The vertical velocity may not be zero near the bed of the estuary as it may be equal to the vertical component of the mean longitudinal current flowing on a sloping bottom, though it will probably tend towards small values there. The vertical velocities are several orders of magnitude smaller than the longitudinal velocities, generally having values of the order 10^{-5} m s^{-1}.

Equation (5.1) can be used in a reduced form to calculate the vertical velocities. If $\bar{u}_A$ is the cross-sectional tidal mean velocity and the breadth b varies with depth and along the estuary, then

$$\frac{\partial b \bar{u}_A}{\partial x} + \frac{\partial b \bar{w}}{\partial z} = 0 \qquad (5.2)$$

The vertical velocities are the mean over an area, parallel to the water surface, between the two cross-sections and the estuary sides, and obviously do not resolve variations across the breadth.

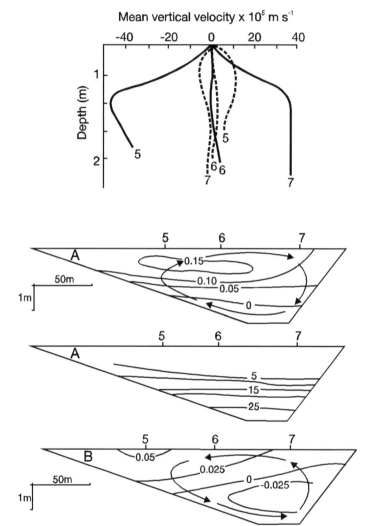

SECONDARY FLOWS

Some care must be taken in interpreting the results of analyses for the vertical velocities. Whereas $\bar{u}$ is the mean longitudinal velocity at a point, or if suitably averaged, the cross-sectional mean, the mean vertical velocity $\bar{w}$ is straightaway the mean averaged over a large area. The local vertical velocities may undergo drastic variation within that area because of the presence of *secondary flows*. These are the velocities in the plane normal to that of the main flow. In estuaries this would be in the cross-section normal to the estuary axis, which itself will change in orientation as the estuary meanders.

In meandering rivers the flow round a bend will experience a *centrifugal force* towards the outside of the bend. This will have the magnitude u^2/r, where r is the radius of curvature of the streamlines. The outward force causes the water and the locus of the maximum current to be thrown towards the outside of the bend, creating a super-elevation of the water surface, with the surface being higher on the outside of the bend. This produces a barotropic force which balances the centrifugal force. Thus $u^2/r = g\delta\zeta/\delta x$.

The balance, though, is only in the depth-averaged conditions. Near the bed friction reduces the velocity and near the surface the velocity is greater. Consequently, there is an imbalance at the surface creating a small net outward flow, whereas near the bed the imbalance is the other way, with a net flow towards the inside of the bend. The secondary circulation is completed by a downward flow on the outside of the bend and an upward flow on the inside, similar to that shown in Figure 5.1A. In a river the outside of a meander bend is usually occupied by a scour hole. Thus, superimposed on the longitudinal flow, is a secondary flow which tends to go downwards into the scour hole and upwards in the shallower water on the inside of the bend. The secondary flow is therefore clockwise on left-handed bends and anticlockwise on right-handed bends, looking downstream. The sense of rotation changes in the straight sections between meanders, where both rotations may be present, but separated in midstream by a shallower longitudinal bar. These secondary flows may be present even in relatively straight channels.

In addition to the effect of the bends there is the effect of changes in the cross-sectional shape along the estuary. The water, as it flows along the

---

Figure 5.1 (*opposite*) Secondary flows in the Vellar estuary. Upper panel: calculated tidal average vertical velocities at cross-section of stations 5–7. Solid lines, high river discharge; broken lines, low river discharge. (A) Interpreted secondary circulation for high river discharge showing (upper) residual longitudinal flow contours ($m\,s^{-1}$) and (lower) mean salinity contours. (B) Secondary circulation for low river discharge showing (upper) residual flow contours ($m\,s^{-1}$) with negative values being flow towards the head and (lower) mean salinity contours. Reproduced by permission of Humana Press from Dyer (1989).

estuary, will both speed up as it encounters shallower water and be deflected towards deeper water. The overall result is a tendency for the flow to spiral as it flows along the estuary.

There is evidence that when a salinity stratification is present the tidally averaged secondary flow system tends to be the opposite of that occurring in rivers. The lateral gradient in salinity creates a baroclinic force which modifies the surface water slopes and the flow structure, such that upward flow occurs in the deeper water where the saline bottom water is thickest, and downwards flow occurs in the shallow water where the saline layer may be absent. Over the deeper water entrainment will also produce a net upward vertical flow, which will be greater than that in shallow water where the stratification will be less.

The contrasting situation between salt wedge and partially mixed conditions in the Vellar estuary is shown in Figure 5.1. For the former conditions the lower saline layer is almost decoupled from the surface layer, but, driven by the shear on the interface, the circulation in the lower layer is opposite to that on the surface. For the latter conditions coupling is larger and the circulation in the lower layer is stronger. Between the two states the inclination of the interface of zero mean velocity appears to have changed its sign, and the difference in the salinity at the surface between the sides of the cross-section was greatest when the sense of the circulation changed. Thus the presence of secondary circulation tends to increase the lateral mixing and consequently the value of the horizontal dispersion coefficient.

In wider estuaries the effect of the rotation of the earth becomes important and the Coriolis force tends to cause the seaward flowing fresher water to hug the right-hand bank in the northern hemisphere and the inflowing saltier water is, likewise, moved towards the opposite bank. This has been cited as the reason for the fresher water in the Chesapeake Bay hugging the western shore, though this is also where the major rivers flow in.

Looking now at the circulations that are likely during the tide, it appears that the same principles hold for estuaries showing a reasonable degree of stratification. However, we will now need to introduce the approximation that $u = \bar{u} + U$. Thus during the tide the centrifugal force $= (\bar{u}^2 + U^2)/r$, and the centrifugal force does not change sign: it is always in the same direction and its magnitude is a non-linear function of the tide velocity, with the mean flow being of only small influence. In contrast, the Coriolis force does change sign during the tide and its magnitude is only linearly related to the tidal velocity. We will consider the relative balance of centrifugal and Coriolis forces further in Chapter 7.

In narrow, well-mixed estuaries the friction produced by the sides becomes important in constraining the flow. On the flood tide the incoming water travels fastest on the surface in the middle of the estuary. It has a higher salinity and consequently would become more dense than the water near the bed and the sides. However, gravity causes it to tend to sink towards the bed as it travels upstream and it draws in water from near the banks as it sinks. On the surface

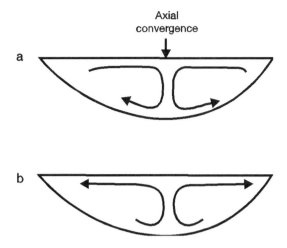

Figure 5.2 Secondary flow in a narrow estuary (a) on the flood tide with surface axial convergence and (b) on the ebb tide with surface divergence.

a convergence develops, with a near-bed divergence, resulting in a secondary circulation produced by two contra-rotating helical flows (Figure 5.2). An *axial convergence* marked by foam and debris is formed that can stretch for many kilometres up the middle of the estuary and is even maintained round the bends. It forms quickly on the flood tide and disappears quickly at high water. This phenomenon has been described by Simpson and Turrell (1986).

On the ebb tide the opposite occurs. The fresher water travels fastest on the surface down the middle of the estuary. Even though this produces a rather more stable situation, mixing creates an upward flow in the middle of the channel. A convergence at the bed and a divergence at the surface are produced.

The question then arises as to what width constitutes a narrow estuary. This has been investigated by Smith (1980) in a modelling study of the dispersion characteristics of channels, which indicates that the longitudinal dispersion coefficient increases with the square of the width until the width reaches about 200 m, and after about 500 m the coefficient decreases. This width is probably the limit at which the secondary circulation changes. In wider estuaries it may also be that during the tide the flow does not actually complete a full cycle of helical flow before the tide turns. Thus the cross-sectional mixing time-scales become long relative to the tide cycle and dispersion becomes reduced.

A further category of longitudinal fronts generated in wide estuaries is often visible on the edges of the channels where shoal water starts. These occur because of differential advection. Observations in the York River, Virginia (Huzzey and Brubaker, 1988) have shown that the water column over the shoals is well mixed, whereas in the channel it is stratified. There is considerable velocity shear between the two water masses due to differences in the amplitude of the tidal velocities with water depth, and there can also be time differences in the turn of the tide. This creates density differences between the shoals and the channel which are greatest at times of minimum currents, and surface convergent flows are then generated by the density differences.

In wide channels dominated by tidal movement there can be marked differences in the strength of ebb and flood currents at different positions in the channel. Again this situation is associated with meanders in the channel and is caused by the flood currents taking a straighter course than the ebb currents, which tend to follow the meanders. These variations produce horizontal circulation, which in the broader parts of the estuary may separate the flows into *ebb and flood channels*. The former contain predominant ebb currents and shallow and narrow towards the sea. The latter narrow and shallow towards the land and contain predominant flood currents. Often these channels occur in pairs, one ending at a steep slope on the edge of the other. As the tidal excursion is generally longer than the length of one of these channels the water flows up one channel and down the other, forming a circulating system. The channels alter their position in an apparently consistent manner and the banks between vary in extent and volume. These movements can cause fluctuations of as much as 5% in estuary volume.

CROSS-SECTIONAL AVERAGING

The measurement of net fluxes through the estuary requires taking measurements at a number of stations and averaging them to represent the full variation across the cross-section. One station in the central channel is unlikely to be representative. This problem is one that is fundamental to most estuarine studies and one which produces potentially large errors. Kjerfve *et al.* (1981) assessed the problem by simultaneously occupying ten stations across the mouth of North Inlet, South Carolina. Their analysis suggested that for a percentage error in the fluxes of $< 15\%$, the lateral station density should be governed by an area of $2 \times 10^6 \, \text{m}^2 \, \text{station}^{-1}$. For most estuaries this means having at least three stations in the cross-section. However, in measurements of current velocity in estuaries we must be careful in the initial positioning of survey stations and cautious about taking cross-sectional averages of currents. Many studies have concentrated measurements in the deep channel, thus only revealing longitudinal and vertical gradients.

Chapter 6

Salt Balance

GENERAL FORMULATION

So far we have examined the distribution of salinity and the velocities in
estuaries and we have made general statements about the principal mechanisms
involved in mixing the salt and fresh water. To predict estuarine characteristics
it is necessary to be able to quantify the water circulation and the mixing
processes. This is done by considering the budget of salt within sections of the
estuary, by adding up the mass of salt being carried into a particular volume
and equating it with what comes out, and the change of salinity within the
volume.

The requirement for this treatment is that salt can be considered to be a
conservative property which can be represented by a continuity equation
similar to that used for calculating the mean vertical velocities. Thus an
accurate knowledge of the distribution of salinity and velocity will be
necessary. Though the results of the analysis will strictly be applicable only
to the salt distribution, it will in fact give information on the probable
distribution of both dissolved and suspended substances that can be considered
to act in a similar way to the salt content.

Consider a small elemental volume of an estuary with sides of length Δx, Δy
and Δz (Figure 6.1). Conservation of salt requires that the net inflow of salt
through the sides in a time Δt will equal the increase of salt content within the
volume. The advective flow of salt through the face $\Delta y \Delta z$ in the time Δt is
$us\Delta y\Delta z\Delta t$. The flow of salt through the opposite face in the same time,
according to Taylor's series, is

$$us\Delta y\Delta z\Delta t + \frac{\partial(us)}{\partial x}\,\Delta x\Delta y\Delta z\Delta t$$

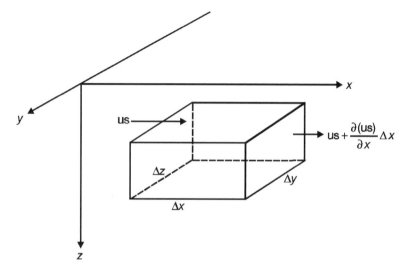

Figure 6.1   Salt transport through an elemental volume.

Consequently, the net inflow of salt in the x direction is

$$- \frac{\partial (us)}{\partial x} \, \Delta x \Delta y \Delta z \Delta t$$

Similarly, in the y and z directions the inflow of salt is

$$- \frac{\partial (vs)}{\partial y} \, \Delta y \Delta x \Delta z \Delta t \quad \text{and} \quad - \frac{\partial (ws)}{\partial z} \, \Delta z \Delta x \Delta y \Delta t$$

The molecular diffusion of salt through the face $\Delta y \Delta z$ will be

$$- \varepsilon \, \frac{\partial s}{\partial x} \, \Delta y \Delta z \Delta t$$

and through the opposite face it will be

$$- \varepsilon \, \frac{\partial s}{\partial x} \, \Delta y \Delta z \Delta t - \frac{\partial}{\partial x} \left(\varepsilon \frac{\partial s}{\partial x} \right) \Delta x \Delta y \Delta z \Delta t$$

where ε is the coefficient of molecular diffusion. The net diffusion in the x direction will thus be

$$+ \frac{\partial}{\partial x} \left(\varepsilon \frac{\partial s}{\partial x} \right) \Delta x \Delta y \Delta z \Delta t = + \varepsilon \, \frac{\partial^2 s}{\partial x^2} \, \Delta x \Delta y \Delta z \Delta t$$

Similarly, the diffusion in the other directions will be

$$+\varepsilon\,\frac{\partial^2 s}{\partial y^2}\,\Delta y\Delta x\Delta z\Delta t \quad \text{and} \quad +\varepsilon\,\frac{\partial^2 s}{\partial z^2}\,\Delta z\Delta x\Delta y\Delta t$$

The amount of salt present in the volume at the time t is $s\Delta x\Delta y\Delta z$. At the time $t + \Delta t$, according to Taylor's series, the amount present will be

$$s\Delta x\Delta y\Delta z + \frac{\partial}{\partial t}\,(s\Delta x\Delta y\Delta z)\Delta t$$

The net increase in salt is

$$\frac{\partial s}{\partial t}\,\Delta x\Delta y\Delta z\Delta t$$

Thus, in the absence of any creation of salt

$$\frac{\partial s}{\partial t} = -\frac{\partial(us)}{\partial x} - \frac{\partial(vs)}{\partial y} - \frac{\partial(ws)}{\partial z} + \varepsilon\left(\frac{\partial^2 s}{\partial x^2} + \frac{\partial^2 s}{\partial y^2} + \frac{\partial^2 s}{\partial z^2}\right) \tag{6.1}$$

This is the equation of continuity for salt for instantaneous values. The continuity equation for volume can be developed in the same way and, considering the fluid to be incompressible, leads to Equation (5.1). Taking the instantaneous salinity and velocities as a tidal mean, a fluctuation of tidal period and a short period turbulent fluctuation: $s = \bar{s} + S + s'$, $u = \bar{u} + U + u'$, $v = \bar{v} + V + v'$ and $w = \bar{w} + W + w'$.

The left-hand side of Equation (6.1), if averaged over a tidal cycle, becomes $\partial\bar{s}/\partial t$ because the tidal fluctuation S will average zero over a tidal cycle, and s' by definition must be zero.

The first term on the right-hand side when multiplied out gives

$$-\frac{\partial}{\partial t}\,(\underset{1}{\bar{u}\,\bar{s}} + \underset{2}{\bar{u}S} + \underset{3}{\bar{u}s'} + \underset{4}{U\bar{s}} + \underset{5}{US} + \underset{6}{Us'} + \underset{7}{u'\bar{s}} + \underset{8}{u'S} + \underset{9}{u's'})$$

and there is a similar set of terms for the lateral and vertical components. Each of these terms can now be averaged over a tidal cycle.

Term 1. The average of this over a tidal cycle is $\bar{u}\,\bar{s}$, the mean advective salt flux.

Terms 2 and 4. There appears to be no reason why the tidal salinity variation should be correlated with the mean velocity or vice versa. Consequently, the tidal mean of these terms is considered negligible.

Terms 6 and 8. As we have already seen, the amplitude of velocity fluctuations increases with tidal current. Consequently, there is a possibility that a correlation may exist between the salinity fluctuations and the tidal velocities and between the velocity fluctuations and the tidal variation of salinity. If we can assume that these correlations do not change over the section of the estuary under consideration, then these terms also will be negligible.

Terms 3 and 7. It is unlikely that the salinity fluctuations will be correlated with the mean velocity and the velocity fluctuations with the mean salinity.

Term 5. If tidal velocity and salinity were both simple trigonometrical functions with a 90° phase difference between them, then the average of their cross-products would be zero. This may be valid in many estuaries because, though $\overline{US}$ may be large, its longitudinal variation $\partial/\partial x(\overline{US})$ may be small. This means that the correlation between the tidal fluctuations of salinity and velocity does not change significantly in the section of the estuary considered.

It is possible to assess the contribution of this term because the observed salinity s and velocity u can be defined as

$$s + \bar{s} + S \quad \text{and} \quad u = \bar{u} + U$$

Consequently,

$$\overline{US} = \overline{(s - \bar{s})(u - \bar{u})} \tag{6.2}$$

where the bar denotes averaging over a tidal cycle.

It is apparent now that if the values are observed over the wrong length of time, then some of the shorter period turbulent fluctuation will appear in the tidal fluctuation cross-product. As the correlations between tidal fluctuations of salinity and velocity can change rapidly in the longitudinal direction, then significant values for $\partial/\partial x(\overline{US})$ can be produced. In the upper reaches of the estuary, where the mean salinity falls off exponentially with distance and where the tidal velocities become markedly disturbed by the river, then $\partial/\partial x(\overline{US})$ may not be negligible. Consequently, in a general treatment it may be advisable to retain this term. Similar arguments about the importance of the time averages of the various lateral and vertical terms can also be used. However, neglecting $\partial/\partial y(\overline{VS})$ and $\partial/\partial z(\overline{WS})$ may again not be valid, though the magnitudes of V and W are relatively small.

Term 9. If turbulent fluctuations of velocity and salinity are correlated, this term will be appreciable even when averaged over a tidal cycle. This is the *Reynolds flux* of salt.

Many analyses neglect the tidal fluctuation cross-products, and the salt balance for values averaged over a tidal cycle and neglecting molecular diffusion will then be given by

$$\frac{\partial \bar{s}}{\partial t} = -\frac{\partial(\bar{u}\bar{s})}{\partial x} - \frac{\partial(\bar{v}\bar{s})}{\partial y} - \frac{\partial(\bar{w}\bar{s})}{\partial z} - \frac{\partial\overline{(u's')}}{\partial x} - \frac{\partial\overline{(v's')}}{\partial y} - \frac{\partial\overline{(w's')}}{\partial z} \tag{6.3}$$

where the bar denotes averaging over a tidal cycle.

In this equation, the *equation of salt continuity*, the first three terms on the right-hand side are the advection terms, the salt flux caused by the mean flow, and the last three are the eddy diffusion terms, the salt flux caused by short period turbulence, which will be much larger than the molecular diffusion. The advective terms involve a mass flux of water as well as salt, whereas the diffusion terms are only associated with a flux of salt.

In general, there is an approximate equilibrium between the terms on the right-hand side of Equation (6.3), so that there is a slow time change of salt content. Under steady-state conditions $\partial \bar{s}/\partial t$ will be zero and the advective and diffusive terms will balance.

Though it is possible to measure the contribution of the advection terms, the eddy diffusion terms are all unknown. Consequently, considerable modification to Equation (6.3) is necessary before it can be used in a realistic situation. The most common modification is made by assuming either that there are no lateral variations, or by dealing with width-averaged values of salinity and velocity. In this case the equation reduces to four terms and solution may be possible in certain circumstances. Other estuaries may be considered to be sectionally homogeneous, in which case the vertical diffusion can be neglected and horizontal diffusion will be the only term to be calculated. However, this is true of very few, if any, estuaries.

The eddy diffusion terms are generally rewritten in a form analogous to that for molecular diffusion, i.e. a constant times the salinity gradient

$$\overline{(u's')} = -K_x \frac{\partial \bar{s}}{\partial x}, \quad \overline{(v's')} = -K_y \frac{\partial \bar{s}}{\partial y} \quad \text{and} \quad \overline{(w's')} = -K_z \frac{\partial \bar{s}}{\partial z} \quad (6.4)$$

Thus the general case for three dimensions [Equation (6.3)] now becomes

$$\frac{\partial \bar{s}}{\partial t} = -\frac{\partial(\bar{u}\bar{s})}{\partial x} - \frac{\partial(\bar{v}\bar{s})}{\partial y} - \frac{\partial(\bar{w}\bar{s})}{\partial z} + \frac{\partial}{\partial x}\left(K_x \frac{\partial \bar{s}}{\partial x}\right) + \frac{\partial}{\partial y}\left(K_y \frac{\partial \bar{s}}{\partial y}\right) + \frac{\partial}{\partial z}\left(K_z \frac{\partial \bar{s}}{\partial z}\right)$$
$$(6.5)$$

This is *Fick's equation*, a classical form of the equation of continuity for salt. The coefficients are called the coefficients of eddy diffusion and have the dimensions of length squared divided by time. These coefficients represent the mixing conditions averaged over a tidal cycle and will have a different physical meaning from the coefficients which would result from using a shorter averaging time of, say, a minute or so (as developed in Chapter 4). Bowden (1963) has termed the coefficients relating to the tidal mean values effective eddy diffusion coefficients.

It is also possible, by using the continuity equation for water [Equation (5.1)] to write Equation (6.5) as

$$\frac{\partial \bar{s}}{\partial t} = \bar{u}\frac{\partial \bar{s}}{\partial x} + \bar{v}\frac{\partial \bar{s}}{\partial y} + \bar{w}\frac{\partial \bar{s}}{\partial z} - \frac{\partial}{\partial x}\left(K_x \frac{\partial \bar{s}}{\partial x}\right) - \frac{\partial}{\partial y}\left(K_y \frac{\partial \bar{s}}{\partial y}\right) - \frac{\partial}{\partial z}\left(K_z \frac{\partial \bar{s}}{\partial z}\right) \quad (6.6)$$

Certain modifications to this general equation are possible if the velocities and salinities can be considered uniform in one or two of the axial directions. These modifications were developed by Pritchard (1958).

For a stratified estuary, an estuary with no spatial variation of salinity in the y direction but with variations in the x and z directions, the continuity equation for volume is

$$\frac{\partial \bar{b}\bar{u}}{\partial x} + \frac{\partial \bar{b}\bar{w}}{\partial z} = 0 \qquad (6.7)$$

where $\bar{b}$ is the mean estuary breadth. The equation for salt continuity is

$$\frac{\partial (\bar{b}\bar{s})}{\partial t} = -\frac{\partial (\bar{b}\bar{u}\bar{s})}{\partial x} - \frac{\partial (\bar{b}\bar{w}\bar{s})}{\partial z} + \frac{\partial}{\partial x}\left(K_x \bar{b}\,\frac{\partial \bar{s}}{\partial x}\right) + \frac{\partial}{\partial z}\left(K_z \bar{b}\,\frac{\partial \bar{s}}{\partial z}\right) \qquad (6.8)$$

Combining Equations (6.7) and (6.8) gives

$$\frac{\partial \bar{s}}{\partial t} = -\bar{u}\,\frac{\partial \bar{s}}{\partial x} - \bar{w}\,\frac{\partial \bar{s}}{\partial z} + \frac{1}{\bar{b}}\,\frac{\partial}{\partial x}\left(K_x \bar{b}\,\frac{\partial \bar{s}}{\partial x}\right) + \frac{1}{\bar{b}}\,\frac{\partial}{\partial z}\left(K_z \bar{b}\,\frac{\partial \bar{s}}{\partial z}\right) \qquad (6.9)$$

For a vertically homogeneous estuary with lateral variation

$$\frac{\partial (\bar{h}\bar{u})}{\partial x} + \frac{\partial (\bar{h}\bar{v})}{\partial y} + \frac{\partial \bar{h}}{\partial t} = 0 \qquad (6.10)$$

and

$$\frac{\partial (\bar{h}\bar{s})}{\partial t} = -\frac{\partial (\bar{h}\bar{u}\bar{s})}{\partial x} - \frac{\partial (\bar{h}\bar{v}\bar{s})}{\partial y} + \frac{\partial}{\partial x}\left(K_x \bar{h}\,\frac{\partial \bar{s}}{\partial x}\right) + \frac{\partial}{\partial y}\left(K_y \bar{h}\,\frac{\partial \bar{s}}{\partial y}\right) \qquad (6.11)$$

Combining Equations (6.10) and (6.11)

$$\frac{\partial \bar{s}}{\partial t} = -\bar{u}\,\frac{\partial \bar{s}}{\partial x} - \bar{v}\,\frac{\partial \bar{s}}{\partial y} + \frac{1}{\bar{h}}\,\frac{\partial}{\partial x}\left(K_x \bar{h}\,\frac{\partial \bar{s}}{\partial x}\right) + \frac{1}{\bar{h}}\,\frac{\partial}{\partial y}\left(K_y \bar{h}\,\frac{\partial \bar{s}}{\partial y}\right) \qquad (6.12)$$

where h is the mean depth.

For a sectionally homogeneous (one-dimensional) estuary

$$\frac{\partial (\bar{A}\bar{u})}{\partial x} + \frac{\partial \bar{A}}{\partial t} = 0 \qquad (6.13)$$

and

$$\frac{\partial (\bar{A}\bar{s})}{\partial t} = -\frac{\partial}{\partial x}\left(\bar{A}\bar{u}\bar{s}\right) + \frac{\partial}{\partial x}\left(\bar{A}K_x\,\frac{\partial \bar{s}}{\partial x}\right) \qquad (6.14)$$

Combining Equations (6.13) and (6.14)

$$\frac{\partial \bar{s}}{\partial t} = -\bar{u}\frac{\partial \bar{s}}{\partial x} + \frac{1}{\bar{A}}\,\frac{\partial}{\partial x}\left(\bar{A}K_x\,\frac{\partial \bar{s}}{\partial x}\right) \qquad (6.15)$$

where $\bar{A}$ is the mean cross-sectional area.

The neglect in Equations (6.11) and (6.14) of the tidal fluctuation cross-products may lead to large errors. In these cases, though $\partial/\partial x\overline{(US)}$ may be small, the presence of a progressive element in the tidal wave may make

$$\frac{\partial}{\partial x}(\overline{hUS}) \quad \text{and} \quad \frac{\partial}{\partial x}(\overline{AUS})$$

significant, where $A = A_0 \cos(\omega t + \theta)$.

Application of Equations (6.7)–(6.15) to estuaries which are not strictly homogeneous in the ways defined, by using width, depth or cross-sectionally averaged values of salinity and velocity, will alter the meaning of the eddy diffusion coefficients. For instance, if Equation (6.12) is used in an estuary where there is in fact stratification, the coefficients K_x and K_y would not represent the mean value of the diffusion coefficients over the vertical. They would become functions which would allow the equation to describe the distribution of the mean salt content, and K_x and K_y would not have the physical significance ascribed to them in Equation (6.4) (Pritchard, 1958).

For the different estuarine types, though, it can be expected that different terms in Equation (6.5) will predominate, other terms being negligible. This is discussed by Pritchard (1955), who reasons that the following salt balances should hold in steady-state conditions.

Salt wedge estuary. The balance is between longitudinal and vertical advection, with the vertical diffusive flux important in the upper layer

$$0 = \bar{u}\,\frac{\partial \bar{s}}{\partial x} + \bar{w}\,\frac{\partial \bar{s}}{\partial z} \qquad (6.16)$$

Partially mixed. With increasing turbulent mixing the vertical diffusion term becomes important throughout the water column

$$0 = \bar{u}\,\frac{\partial \bar{s}}{\partial x} + \bar{w}\,\frac{\partial \bar{s}}{\partial z} - \frac{\partial}{\partial z}\left(K_z\,\frac{\partial \bar{s}}{\partial z}\right) \qquad (6.17)$$

Vertically homogeneous with lateral variations. The vertical terms are now unimportant and lateral advection and diffusion are large

$$0 = \bar{u}\,\frac{\partial \bar{s}}{\partial x} + \bar{v}\,\frac{\partial \bar{s}}{\partial y} - \frac{\partial}{\partial y}\left(K_y\,\frac{\partial \bar{s}}{\partial y}\right) \qquad (6.18)$$

Sectionally homogeneous. The balance is entirely between longitudinal advection and diffusion

$$0 = \bar{u}\,\frac{\partial \bar{s}}{\partial x} - \frac{\partial}{\partial x}\left(K_x\,\frac{\partial \bar{s}}{\partial x}\right) \qquad (6.19)$$

In *fjords*, the mixing characteristics will depend largely on the depth of the sill at the mouth. For a deep sill the mixing characteristics are likely to be similar to the partially mixed type, but with the bottom layer replaced with a basin of undiluted sea water. For a shallower sill depth the mixing is more likely to be similar to the salt wedge estuary.

We shall now consider some examples of the applications that have been made of the equation of salt continuity. This should illustrate the validity of using reduced forms of Equation (6.5) in different types of estuary.

KNUDSEN'S HYDROGRAPHICAL THEOREM

If we can consider the mean flow as occurring in two layers and if we can assume that diffusion is negligible, then Knudsen's hydrographical theorem can be developed from consideration of salt and volume continuity under steady-state conditions.

If A denotes the cross-sectional area of each layer in Figure 6.2, then volume continuity requires that

$$A_1 u_1 - A_3 u_3 = A_2 u_2 - A_4 u_4$$

and continuity of salt requires that

$$A_1 u_1 S_1 - A_3 u_3 S_3 - A_2 u_2 S_2 + A_4 u_4 S_4 = 0$$

For the steady-state the net rate of transport of salt across each section is zero, then

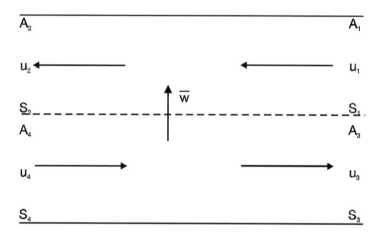

Figure 6.2 Longitudinal section through an estuary showing notation for Knudsen's hydrographical theorem.

$$A_1 u_1 S_1 = A_3 u_3 S_3 \quad \text{and} \quad A_2 u_2 S_2 = A_4 u_4 S_4$$

Thus

$$A_2 u_2 \left(1 - \frac{S_2}{S_4} \right) = A_1 u_1 \left(1 - \frac{S_1}{S_3} \right)$$

and

$$A_4 u_4 \left(\frac{S_4}{S_2} - 1 \right) = A_3 u_3 \left(\frac{S_3}{S_1} - 1 \right)$$

If the section comprising A_1 and A_3 is the head of the estuary and R is the river flow

$$A_1 u_1 - A_3 u_3 = R \quad \text{and} \quad A_1 u_1 S_1 = A_3 u_3 S_3 = 0$$

Consequently, $\quad R = A_2 u_2 - A_4 u_4 \quad$ and $\quad A_2 u_2 S_2 = A_4 u_4 S_4$

Thus

$$A_2 u_2 = \frac{R S_4}{S_4 - S_2} \quad \text{and} \quad A_4 u_4 = \frac{R S_2}{S_4 - S_2} \tag{6.20}$$

Between the two sections vertical advection completes the circuit and

$$\overline{w} B = A_4 u_4 = R S_4 / S_4 - S_2 \tag{6.21}$$

where $\overline{w}$ is the mean vertical velocity and B is the surface area of the interface between the two sections.

This approach is only valid for a purely advective salt balance and consequently may only be valid in salt wedge estuaries, fjords and straits. The hydrographical theorem is not a different concept, but only a formulation of a special case of the generalized salt balance equation.

PARTIALLY MIXED ESTUARIES

If we assume, following Pritchard's work in the James River, that the horizontal advection and vertical eddy diffusion are the dominant processes affecting the distribution of salinity, and that this is applicable to observed values at any depth z, as well as mean values, then estimates of the vertical eddy diffusion can be made by a method used by Bowden (1960, 1963).

For the salt balance for an element at depth z, Equation (6.17) becomes

$$\frac{\partial s_z}{\partial t} + u_z \frac{\partial s_z}{\partial x} = \frac{\partial}{\partial z} \left(K_z \frac{\partial s_z}{\partial z} \right) \tag{6.22}$$

Continuity requires that

$$\frac{\partial \langle s_z \rangle}{\partial t} + \left\langle u_z \frac{\partial s_z}{\partial x} \right\rangle = 0$$

where the brackets $\langle \ \rangle$ denote a depth mean taken from the surface to the bottom. The salinity at any depth can be considered as a depth mean and a deviation from the depth mean, i.e. $s_z = \langle s \rangle + s_V$. Then in the case of steady flow, at any depth

$$\frac{\partial s_V}{\partial t} + u_z \frac{\partial s_z}{\partial x} - \left\langle u_z \frac{\partial s_z}{\partial x} \right\rangle = \frac{\partial}{\partial z}\left(K_z \frac{\partial s_z}{\partial z} \right) \tag{6.23}$$

With zero boundary conditions at the surface, Equation (6.23) can be integrated stepwise from the surface to the bottom and the distribution of K_z calculated. The equation can also be integrated over a tidal cycle, but the vertical eddy diffusion coefficient will not then bear close relationship to the shorter term values, as it will represent the average of the mixing conditions over the tidal cycle. Values of this 'effective' coefficient given by Bowden (1963) and Bowden and Sharaf el Din (1966a) are shown in Table 6.1.

In an attempt to investigate the temporal variation of the vertical eddy diffusion coefficient, Bowden (1963) applied Equation (6.23) to hourly data, but, because of measurement errors, consistent results were obtained for only one period. These indicated that when the current was large the values of K_z were three to five times the effective values over the tidal period. The maximum value was $155\,\mathrm{cm^2\,s^{-1}}$ at mid-depth three hours after high water. These values, however, were still smaller than those which one would expect in conditions of neutral stability.

Table 6.1   Coefficient of vertical eddy diffusion K_z ($\mathrm{cm^2\,s^{-1}}$) at various points in the Mersey Narrows (for station positions, see Figure 9.5).

	Station number						
z/h	C2	C2	E2	C2	C1	C2	C3
0.1	5.5	9	5.5	2.8	8	6	5
0.3	18	24	27	9	12	15	13
0.5	28	33	71	28	22	28	17
0.7	23	34	41	28	22	23	9
0.9	8	16	26	15	1	3	1

WELL-MIXED ESTUARIES

In one-dimensional estuaries, where the distribution of properties can be considered dependent on x only, considerable use has been made of Equation (6.14) integrated once with respect to x

$$\overline{us} = K_x \frac{\partial \bar{s}}{\partial x} \qquad (6.24)$$

This states that the longitudinal advection of salt downstream on the sectional mean flow is balanced by an upstream horizontal diffusion. The sectional mean velocity is related to the river flow if $\bar{u} = R/\overline{A}$.

Therefore

$$K_x = R\bar{s} \Big/ \overline{A} \frac{\partial \bar{s}}{\partial x} \qquad (6.25)$$

Thus K_x can be calculated for any position in an estuary if the river flow, the cross-sectional area and the distribution of salinity along the estuary are known. Some values are shown in Table 6.2. It is noticeable from Equation (6.25) how K_x decreases linearly with river flow, other things being equal.

This approach has been used by West and Williams (1972) in the Tay estuary, Scotland, to compare the value of K_x with river discharge at various points along the estuary. The results (Figure 6.3) show that Equation (6.25) holds at the mouth for all river flows. At points further up the estuary it holds for lower flows, but near the head for only very low flows. This is due to the increased river flow pushing the stratification gradually downstream, which makes the equation strictly inapplicable.

This approach takes no account of lateral or vertical variations in the mean flow, however. The effect of any gravitational circulation on the salt transport will appear in the diffusive term, in spite of not being the result of a turbulent exchange. Consequently, the coefficient is really a dispersion coefficient. In Equation (6.25) the experimentally observed upstream salt flux will be due to a variety of processes, including gravitational convection, and K_x is thus a *coefficient of longitudinal dispersion*, or a coefficient of effective longitudinal diffusivity.

CROSS-SECTIONAL SALT FLUXES

An alternative approach to the salt balance in an estuary is to consider the cross-sectionally averaged salt fluxes during the tide. This requires measuring the velocities and the salinity at a number of positions across the cross-section. The salt flux is then their product. This approach considers the contributions of the depth mean product and that resulting from the product of the deviations

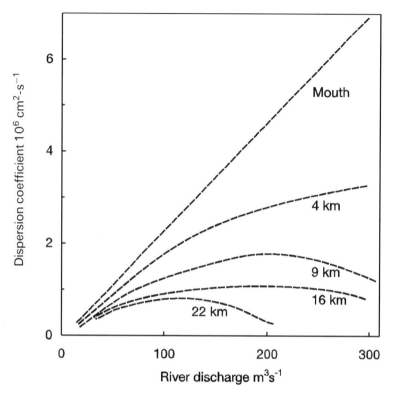

Figure 6.3 Distribution of longitudinal dispersion coefficient at different distances from the mouth for different river flows in the Tay estuary. Reproduced from West, J.R. and Williams, D.J.A. (1972) An evaluation of mixing in the Tay estuary. *Proc. 13th Coastal Eng. Conf.*, 2153–2169, by kind permission of the American Society of Civil Engineers.

of velocity and salinity from their depth means. Most of the latter is the result of velocity shear on the estuary bed and gravitational circulation.

The effect of vertical shear has been examined by Bowden (1963). An estuary with vertical, but without lateral, variations is considered. The instantaneous rate of transport of salt through a unit width of a section perpendicular to the mean flow is given by

$$Q = \int_0^h us \; dz = h\langle us \rangle \qquad (6.26)$$

where h is the depth. At any depth let $u = u_z + u'$ and $s = s_z + s'$, where u_z and s_z are the observed values averaged over a minute or so, and u' and s' are the irregular, turbulent variations.

u_z and s_z may also be written as

$$u_z = \langle u \rangle + u_V \quad \text{and} \quad s_z = \langle s \rangle + s_V$$

where

$$\langle u \rangle = \frac{1}{h} \int_0^h u_z \, dz \quad \text{and} \quad \langle s \rangle = \frac{1}{h} \int_0^h s_z \, dz$$

Thus u_V and s_V are the deviations with depth from the mean values over the entire depth $\langle u \rangle$ and $\langle s \rangle$.

Because of tidal fluctuations $\langle u \rangle$ and $\langle s \rangle$ will vary regularly over a tidal cycle. Consequently $\langle u \rangle = \bar{u} + U$ and $\langle s \rangle = \bar{s} + S$, where

$$\bar{u} = \frac{1}{T} \int_0^T \langle u \rangle \, dt \quad \text{and} \quad \bar{s} = \frac{1}{T} \int_0^T \langle s \rangle \, dt$$

so that $\bar{u}$ and $\bar{s}$ are the mean values of $\langle u \rangle$ and $\langle s \rangle$ over the tidal cycle.

Consequently, Equation (6.26) becomes, averaged over a tidal cycle

$$\bar{Q} = \overline{h\bar{u}\bar{s}} + \frac{1}{T} \int_0^T hUS \, dt + \frac{1}{T} \int_0^T h\langle u_V s_V \rangle \, dt + \frac{1}{T} \int_0^T h\langle u's' \rangle \, dt \qquad (6.27)$$

or

$$\bar{Q} = \overline{h\bar{u}\bar{s}} + \overline{(hUS)} + \overline{h\langle u_V s_V \rangle} + \overline{h\langle u's' \rangle}$$

$$= Q_1 + Q_2 + Q_3 + Q_4$$

where the brackets $\langle \ \rangle$ denote the depth means of the products $u_V s_V$ and $u's'$ and $\bar{h}$ is the mean depth.

The first term Q_1 represents the contribution to the total salt flux by the Eulerian mean flow, which will not be equal to the river discharge when Stokes drift is important. Q_2 arises from the variation in the depth mean velocity and salinity during the tidal period. If the tidal fluctuations are harmonic and 90° out of phase then Q_2 should be zero, providing the tide acts purely as a standing wave. It will provide a finite negative amount if the water flowing through the section on the flood has a higher salinity than that flowing seaward on the ebb. This could occur if the tide does not act as a purely standing wave. The process described by this term is generally known as *tidal pumping*. Q_3 arises from the variation of the velocity and salinity with depth, the shear effect, and will be finite if the two are correlated. Thus a positive (downstream) u_1 associated with a negative s_1, a lower salinity, will produce a finite Q_3. This could also be termed a gravitational convection mode and includes the effect of vertical diffusion. Q_4 represents the salt flux on the short period turbulence, which is normally considered small enough in comparison with the other terms to be negligible.

Equation (6.27) can be written in the form

$$\overline{Q} = \overline{hu}\,\overline{s} - K_x \overline{h}\,\frac{\partial \overline{s}}{\partial x} \tag{6.28}$$

The coefficient of longitudinal dispersion includes the effect of Q_2 and Q_3 as well as the short period turbulent term Q_4. Consequently, this coefficient will alter with magnitude of the vertical eddy diffusion. Bowden (1965) has shown that K_x, calculated this way, is inversely proportional to K_z so that a decrease in vertical diffusion due to, for instance, a stable density gradient, gives rise to an increase in the longitudinal dispersion.

From data obtained in the Mersey Narrows, Bowden (1963) has calculated the values of Q_1, Q_2 and Q_3. From the observed data Q_4 cannot be calculated, but it is assumed to be small compared with the other terms. Q_1 representing the advection on the mean flow, is always downstream, but is more than compensated by the upstream salt flux of Q_2 and Q_3. Thus there is a net upstream flux of salt, small at the second period, but large during the other three. This may be the result of one or several of the following causes:

Table 6.2 Longitudinal eddy diffusion coefficient K_x for some estuaries calculated using Equation (6.25).

	Vertical salinity difference	K_x $(10^5 \text{ cm}^2 \text{ s}^{-1})$
Severn (summer)		
Aust	0.3	17.4
Portishead	0.1	10.6
Severn (winter)		
Aust	0.5	30.9
Portishead	0.3	15.7
Thames (low flow)		
10 miles below London Bridge	1.0	5.3
25 miles below London Bridge	1.0	8.4
Mersey Narrows		
Low flow	1.3	16.1
High flow	1.5	36.0
Southampton Water	1.2	15.8

1. Horizontal diffusion on the short period turbulence (Q_4), which has been neglected.
2. A time change in mean salinity, so that $\partial \bar{s}/\partial t \neq 0$, i.e. $\overline{Q} \neq 0$. This could be as a result of changing weather effects or river discharge.
3. Lateral variations across the estuary especially in velocity, caused by secondary currents.
4. Measurement errors.

This approach has been widely used despite the problems involved in obtaining a valid measure of the salt fluxes within the estuary from it. Most workers, however, consider the deviations from the cross-sectional averages rather than from the depth mean values. This then considers the lateral variations in current velocity and salinity arising from transverse shear, but requires an extension of the above analysis.

Okubo (1964) has considered rigorously the averaging processes and their validity when applied to the mass and pollutant balance on a cross-section of an estuary. The instantaneous salt balance equation can be written as

$$\frac{\partial(As_A)}{\partial t} = -\frac{\partial}{\partial x}[As_A u_A + A(s_d u_d)_A] \tag{6.29}$$

where the instantaneous salinity and velocity is considered as being composed of a cross-sectional mean and a deviation from this mean, i.e. $s = s_A + s_d$ and $u = u_A + u_d$. The subscript A denotes averaging over the cross-sectional area. The instantaneous values are then considered as a tidal average and a short period fluctuation. Thus

$$A = \overline{A} + A', \quad s_A = \bar{s}_A + s'_A, \quad u_A = \bar{u}_A + u'_A$$

$$\text{and} \quad (s_d u_d)_A = \overline{(s_d u_d)_A} + (s_d u_d)'_A$$

Substituting into Equation (6.29), averaging over a tidal cycle and eliminating terms leads to

$$\frac{\partial}{\partial t}(\overline{A}\bar{s}_A) = -\frac{\partial}{\partial x}[\bar{s}_A(\overline{A}\bar{u}_A + \overline{A'u'_A}) + \overline{A}(\overline{s'_A u'_A} + \overline{(s_d u_d)_A})] \tag{6.30}$$

Hansen (1965) has used the same approach, but considered each component as the sum of a tidal mean, a tidal oscillation and a turbulent fluctuation, all averaged over the cross-section. Thus

$$A = \overline{A} + A + A', \quad u_A = \bar{u}_A + U_A + u'_A, \quad s_A = \bar{s}_A + S_A + s'_A$$

$$\text{and} \quad (u_d s_d)_A = \overline{(u_d s_d)_A} + (U_d S_d)_A + (u_d s_d)'_A$$

The mean salt flux over a tidal cycle through a cross-section is

$$\overline{F}_s = \overline{Au_A s_A} + \overline{A(u_d s_d)_A} \tag{6.31}$$

Thus

$$\overline{F}_s = \overline{(\overline{A} + A + A')(\overline{u}_A + U_A + u'_A)(\overline{s}_A + S_A + s'_A)}$$

$$+ \overline{(\overline{A} + A + A')((\overline{u_d s_d})_A + (U_d S_d)_A + (u_d s_d)'_A)}$$

The mean flux of water through the section during the tide is

$$R = \overline{A \overline{u}_A} + \overline{A U_A} + \overline{A' u'_A}$$

This is comparable with Equation (4.15) as the third term is likely to be extremely small.

Multiplying out and eliminating terms gives

$$\overline{F}_s = R\overline{s}_A + \overline{A}\,\overline{U_A S_A} + \overline{u}_A \overline{A S_A} + \overline{A U_A S_A} + \overline{A}\,\overline{(u_d s_d)_A} + \overline{A(U_d S_d)_A}$$

$$+ (\overline{A}\,\overline{u'_A s'_A} + \overline{u}_A \overline{A' s'_A} + \overline{A' u'_A s'_A} + \overline{A'(u_d s_d)'_A}) \qquad (6.32)$$

The terms on the right side of Equation (6.32) express the seaward salt flux associated with

1. The mean river flow and salinity.
2. Correlation of tidal variations of the sectional mean salinity and current.
3. Correlation between tidal variations of cross-section and mean salinity.
4. Third-order correlation of tidal variations in mean salinity, velocity and cross-section.
5. Mean shear effect produced by gravitational circulation.
6. Covariance of tidal shear effect and cross-section.
7. Correlation between short period fluctuations.

Terms 2 and 5 are similar to Q_2 and Q_3 of Equation (6.27). Terms 2–4 are called the tidal pumping terms. Only the first term in the brackets of term 7 will be of any significance.

As an example we will take analysis of observations from the Columbia River (Hansen, 1965). The observations gave equations showing the amplitudes and phases of $\overline{A} + A$, $\overline{s}_A + S_A$, $\overline{u}_A + U_A$ and $\overline{(u_d s_d)} + (U_d S_d)_A$. These equations enabled calculation of terms 1–6 of Equation (6.32). The change in salt storage above the section was assumed to be small and the results showed that, of the salt advected seawards in the mean river discharge, about 40% was balanced by covariance between fluctuations of tidal period (term 2) and about 45% was balanced by shear effects (term 5). The remaining 15% was considered to be caused by the short period fluctuations (term 7) as the other terms were small and were comparable with the uncertainty in estimation of the major fluxes. Thus the tidal pumping and the gravitational circulation were about equal.

From volume continuity considerations it was calculated that 70% of the non-tidal drift was caused by river discharge and 30% was compensation for Stokes' drift as a result of the effect of a progressive component on the tidal wave. The salinity variation had a phase lag of 27° compared with the cross-sectional area, and the maximum currents occurred 36° after the mean area. These lags produced the large value for $\overline{A U_A}$, and the large values for $\overline{A U_A S_A}$. However, the mean product $\overline{A U_A S_A}$ is small. Consequently, the progressive component of the tidal change in cross-sectional area is cancelling out the effect that the tidal fluctuations of salinity and velocity produce by not being 90° out of phase.

Transverse Shear

It would be instructive to analyse the shear effect in such a way that it is possible to separate the effects of vertical shear and transverse shear. Their comparative importance may illustrate better the processes operating to cause a balance of salt in estuaries.

If u_T is the deviation of the depth mean velocity $\langle u \rangle$ from a cross-sectional mean velocity u_A and u_A has a tidal fluctuation given by $u_A = \bar{u}_A + U_A$ then the instantaneous velocity u can be represented by

$$u = \bar{u}_A + u_V + u_T + U_A + u'$$

where u_V is the deviation of the observed values from the depth mean. The salinity variations can be represented in a similar way and the cross-product averaged over a tidal cycle will be

$$\overline{us} = \bar{u}_A \bar{s}_A + \overline{u_V s_V} + \overline{u_T s_T} + \overline{U_A S_A} + \overline{u's'}$$

No variation of breadth b, or of depth h during the tide is considered. If $\bar{u}_A = R/\bar{A}$ then integration over the cross-section will give the mean salt flux as

$$\bar{F}_s = R\bar{s}_A + b \int_0^h \overline{u_V s_V}\, dz + h \int_0^b \overline{u_T s_T}\, db + A\overline{U_A S_A} - A\overline{(u's')}_A$$

In steady-state conditions this becomes

$$0 = R\bar{s}_A + b \int_0^h \overline{u_V s_V}\, dz + h \int_0^b \overline{u_T s_T}\, db + A\overline{U_A s_A} - A K_x \frac{\partial \bar{s}_A}{\partial x} \qquad (6.33)$$

$$\quad\; 1 \qquad\qquad 2 \qquad\qquad\qquad 3 \qquad\qquad 4 \qquad\quad 5$$

The terms on the right-hand side of Equation (6.33) represent the salt flux associated with

1. The mean river flow.
2. The vertical gravitational circulation.
3. The transverse circulation.
4. Correlation of tidal fluctuations of velocity and salinity, tidal pumping.
5. Diffusion on the short period turbulence.

In the event of several stations on a cross-section being occupied simultaneously for a sufficient time, only the term 5 will be unknown and K_x can be calculated.

Bowden and Gilligan (1971) have regarded term 3 as part of the turbulent diffusion term (T_D) with a suitably defined modified dispersion coefficient, and have assumed the tidal fluctuation cross-product (term 4) as negligible. Thus Equation (6.33) becomes

$$0 = R\bar{s}_A + b \int_0^h \overline{u_V s_V}\, \mathrm{d}z - AK_x' \frac{\partial \bar{s}_A}{\partial x} \qquad (6.34)$$

or

$$0 = T_R - T_A - T_D$$

This expresses the downstream salt transport by the river discharge T_R as balanced by T_A, the advective upstream transport by the density current, the net vertical circulation and by upstream diffusion, which includes the effect of the net transverse circulation. Consequently the *diffusive fraction of the total upstream salt flux* ν (see also Chapter 9) is given by

$$\nu = \frac{T_D}{T_A + T_D} = \frac{T_R - T_A}{T_R} \qquad (6.35)$$

Strictly, the diffusive fraction should be termed the dispersive fraction as there are more processes included in it than turbulence. Both definitions are used.

Values for T_R and T_A were computed directly for data from the Mersey. The calculated diffusive fraction ν was least in the centre part of the Narrows, where the density current was most highly developed. The fraction increased towards either end of the Narrows where the estuary widens and becomes shallower. At the Liverpool Bay end, where the exchange of salt is probably carried out largely by horizontal eddies, the diffusive fraction was 0.85. In the Columbia River estuary $\nu = 0.67$ at high discharge and 0.59 at low discharge (Hughes and Rattray, 1980).

It is interesting to note that in this analysis the upstream salt flux by the density current (T_A) is advective, whereas in the previous method [Equation (6.28)] this term would be included in the diffusive part. This illustrates how the values obtained for the longitudinal dispersion coefficient depend on the method of analysis of the data.

In Equation (6.33) the effect of cross-sectional deviation of the tidal fluctuations of velocity and salinity from the cross-sectional average is not considered. Thus term 4 could be separated into two parts, one expressing the transverse and the other the vertical variation in the tidal fluctuations. This is done in the analysis of Fischer (1972). However, in this analysis the effect of fluctuations in cross-sectional area is not considered.

The observed velocity

$$u_z = u_A + u_d$$

The cross-sectional mean velocity undergoes a tidal fluctuation and can be represented by $u_A = \bar{u}_A + U_A$. Similarly, $u_d = \bar{u}_d + U_d$. Considering a three-dimensional profile both $\bar{u}_d$ and U_d can be separated into variations in the vertical and transverse directions. Thus $\bar{u}_d = \bar{u}_T + \bar{u}_V$ and $U_d = U_T + U_V$. Salinity can be treated similarly. The mean salt flux over a tidal cycle through a unit area of the cross-section is then:

$$\bar{F}_S = \bar{u}_A \bar{s}_A + \overline{U_A S_A} + (\overline{\bar{u}_T \bar{s}_T})_A + (\overline{\bar{u}_V \bar{s}_V}) + \overline{(U_T S_T)_A} + \overline{(U_V S_V)_A} \qquad (6.36)$$

In Equation (6.36) the terms on the right-hand side represent the salt flux due to

1. Mean flow on the river discharge.
2. Correlation of tidal fluctuations of sectional mean salinity and velocity.
3. Net transverse circulation.
4. Net vertical circulation.
5. Transverse oscillatory shear.
6. Vertical oscillatory shear.

Fischer (1972) evaluated the contribution of the last four terms to the dispersion coefficient, from data for the Mersey. He concluded that the net transverse circulation was dominant, with the net vertical circulation and the vertical oscillatory terms an order of magnitude smaller. These results suggest that the net vertical circulation appears to be less important than the analysis of Bowden and Gilligan (1971). They calculated that about half of the longitudinal transport of salt was due to the net vertical circulation.

If we include the effect of tidal variation in cross-sectional area, this would lead to Equation (6.37).

Using the definitions

$$A = \bar{A} + A + A', u_A = \bar{u}_A + U_A + u'_A,$$
$$s_A = \bar{s}_A + S_A + s'_A, u_d = \bar{u}_d + U_d + u'_d, s_d = \bar{s}_d + S_d + s'_d$$

where

$$\bar{u}_d = \bar{u}_T + \bar{u}_V, \bar{s}_d = \bar{s}_T + \bar{s}_V \quad \text{and} \quad U_d = U_T + U_V, S_d = S_T + S_V$$

$$\text{and} \quad u'_d = u'_T + u'_V, s'_d = s'_T + s'_V$$

Then the mean flux of salt through a section over a tidal cycle (Equation 6.29) becomes, neglecting some terms

$$\boldsymbol{F}_S = \overline{A\bar{u}_A\bar{s}_A} + \overline{A U_A \bar{s}_A} + \overline{A\,U_A S_A} + \overline{A U_A S_A} + \overline{A(\bar{u}_T\bar{s}_T)_A} + \overline{A(\bar{u}_V\bar{s}_V)_A}$$

$$+ \overline{A\,(U_T S_T)_A} + \overline{A\,(U_V S_V)_A} + \overline{A(U_T S_T)_A} + \overline{A(U_V S_V)_A} + \overline{A\,(u'_T s'_T)_A} + \overline{A\,(u'_V s'_V)_A}$$

$$(6.37)$$

Of the terms on the right-hand side, Hansen (1965) found that term 3 and terms 5 and 6 were important, and Fischer (1972) found term 5 the most important. Though terms 11 and 12 are those relating to eddy diffusion, all terms except the first would be included in the calculation of a longitudinal dispersion coefficient according to Equation (6.28). Thus in comparison with Equation 6.35, terms 1 and 2 would be T_R, terms 5 and 6 would be T_A and the rest would be T_D.

Dyer (1974) used data from three estuaries and showed that for the Vellar estuary vertical gravitational circulation was the largest of the shear terms, but for the partially mixed Southampton Water and Mersey estuaries the lateral contribution was of the same order as the vertical. Dividing the sum of terms 7–12 by the mean cross-sectional area times the longitudinal salinity gradient gave estimates of the dispersion coefficients of 20 to over $300\,\mathrm{m}^2\,\mathrm{s}^{-1}$. Murray and Siripong (1978) found that the lateral terms were 1.5 times the vertical gravitational circulation in the estuary of the Rio Guayas, Ecuador. However, Rattray and Dworski (1980) have shown that the results of the analyses are sensitive to the sampling design and whether the increments of cross-sectional area are conformal to the bottom or to the water surface. This becomes important during separation of the transverse and vertical deviations. They showed that, for Southampton Water, the gravitational circulation accounted for all of the shear transport. Hughes and Rattray (1980) applied an improved formulation to the highly stratified Columbia River estuary and found that tidal pumping and the vertical gravitational circulation were the largest terms. The approach was also applied to a time series of 31 tidal cycles observed in a Malaysian mangrove estuary by Dyer *et al.* (1992). They found that there was a dynamic coupling between the transverse and vertical contributions. During a river discharge event there was an outflow of salt associated with the vertical oscillatory shear until the vertical gravitational circulation became fully developed. As river flow diminished an outflow of salt occurred on the vertical oscillatory shear as the gravitational circulation diminished. There was a time lag of two to three tides in the dynamic response to the flow changes. The magnitude of the transverse mean shear term was about half that of the vertical mean shear term. The vertical gravitational circulation diminished rapidly as the estuary destratified towards spring tides. Dispersion coefficients had values of $340 \, \mathrm{m^2 \, s^{-1}}$ during the discharge event, and $100 \, \mathrm{m^2 \, s^{-1}}$ and $140 \, \mathrm{m^2 \, s^{-1}}$ for neap and spring tides, respectively.

Lewis (1979) examined the variations in salt flux at stations on several cross-sections of the Tees estuary. The local topographic features, such as bends, channels and jetties, had considerable influence on the transverse deviations from the cross-sectional means of the steady and tidal flows, and hence the salt fluxes. However, the transverse variation in the vertical gravitational circulation was little affected by the topography, but it was likely to be more affected by changes in river flow. The gravitational circulation term was the same magnitude as the tidal pumping, which also varied across the estuary.

However, Lewis and Lewis (1983) also found for the Tees that the transverse variations were of secondary importance. Their analysis, though basically the same as that above, considers the amplitude and the phase angle between the different variables. Along the estuary there were variations in the relative contributions of the different terms. As the estuary narrows, the oscillatory contributions dominate over that due to shear. In the reach of the estuary where the bulk Richardson numbers were highest, they found that the variations in phase of the tidal currents over depth led to a reinforcement of the gravitational circulation. Nevertheless, this effect contributed less than a quarter of the observed vertical circulation.

Uncles *et al.* (1985a, 1985b) have used basically the same formulation as Equation (6.27), but with different symbols, for seven stations in the Tamar estuary at spring and neap tides and medium to high river flows. Stokes drift was only important in the shallow upper reaches. Both tidal pumping and gravitational circulation were directed up-estuary at spring tides, with the former being dominant. Tidal pumping was also directed up-estuary at neap tides, but magnitudes were smaller and vertical shear then dominated in the lower reaches. There was a characteristic transverse structure. The gravitational circulation was stronger in the central deeper parts of the cross-section and tidal pumping was directed up-estuary in the deep parts, but sometimes reversed in the shallow areas.

Consequently, gravitational circulation and tidal pumping are both important, but with their relative importance varying with the estuary type, river flow and tidal range. There are also significant differences across the estuary so that the values for the central channel are not representative of the entire cross-section.

Many of the studies have shown that there is a considerable discrepancy between the sum of the pumping terms and the gravitational circulation with the gauged river flow and the mass of salt within the estuary. These seem to be too great to be a time mean accumulation or removal of salt within the estuary. In other words, the estuary would have to be in a considerably non-steady state. One explanation for this may be the result of unacceptable weighting of the fluxes by having non-constant incremental areas, and because they change in number or size during the tide. Some of the analyses have been carried out with rectangular increments of constant area, though this makes accurate representation of the cross-section difficult. A further problem is obtaining good measurements in the shallow parts of the cross-section.

Considerable care needs to be exercised in the application of the methods for separating the various contributions to the cross-sectional salt flux. The methods are of more use in relative than in absolute terms, but the division of lateral and vertical contributions poses considerable problems.

Chapter 7

Dynamic Balance

GENERAL FORMULATION

In the last chapter we examined the salt balance in an estuary and the water flows required to produce this balance. There are other constraints, however, because the water flows, as well as satisfying the salt balance, must also be acting according to a dynamic balance. The density distribution produces pressure gradients and these are a major force controlling the current flows. Turbulent eddies produced by the flows cause exchanges of momentum, as well as of salt, and produce frictional forces helping to resist the flow.

The dynamic balance is provided by *Newton's second law of motion*, which states that force equals mass times acceleration. The forces involved should include the sea surface gradients and internal density gradients which give horizontal differences in pressure, the Coriolis force due to the earth's rotation, wind stresses on the sea surface and frictional forces on the sea bed, turbulent stresses or eddy viscosity and molecular viscosity. The accelerations will be comprised of terms involving change in velocity with time, or change in velocity in space, of the form $\partial u / \partial t$ and $u \partial u / \partial x$.

Consider an element with sides of lengths Δx Δy and Δz (Figure 7.1) in a coordinate system where the x axis is positive down the estuary, the y axis is directed across the estuary, positive to the right, and the z axis is vertically downwards. The forces acting on the element are composed of surface forces and body forces. The surface forces consist of two parts, one acting parallel to the surfaces of the element (shear) and the other acting normal to the surface (pressure). Body forces depend on the mass of fluid in the element and are caused by gravity and the Coriolis effect.

The shearing stress is the means by which the water above a horizontal plane acts on that below. The frictional forces are derived from the vertical changes in the horizontal shearing stress. If the stress is constant with depth there is no

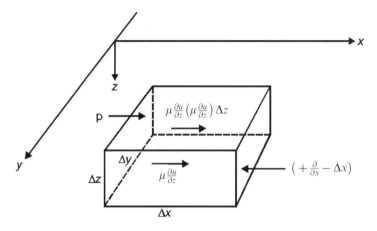

Figure 7.1 Forces acting on an element.

frictional force and if the stress varies linearly with depth, then the frictional force is constant.

The frictional force on the bottom of the element in the x direction at any instant is

$$\mu \frac{\partial u}{\partial z} \Delta x \Delta y$$

where, μ is the coefficient of molecular viscosity. On the top of the element, by Taylor's series, the frictional force is

$$\mu \frac{\partial u}{\partial z} \Delta x \Delta y + \frac{\partial}{\partial z}\left(\mu \frac{\partial u}{\partial z}\right) \Delta z \Delta x \Delta y$$

This component of the total frictional force is the difference

$$\frac{\partial}{\partial z}\left(\mu \frac{\partial u}{\partial z}\right) \Delta z \Delta x \Delta y = \mu \frac{\partial^2 u}{\partial z^2} \Delta z \Delta x \Delta y$$

Similarly, the component of the total frictional force at any instant caused by friction on the sides of the element will be

$$\mu \frac{\partial^2 u}{\partial y^2} \Delta y \Delta x \Delta z$$

and on the ends of the element

$$\mu \frac{\partial^2 u}{\partial x^2} \Delta x \Delta y \Delta z$$

The net pressure force on the element in the x direction is

$$p\Delta y\Delta z - \left(p + \frac{\partial p}{\partial x}\right)\Delta y\Delta z = -\frac{\partial p}{\partial x}\Delta x\Delta y\Delta z$$

In the x direction the Coriolis force is $(2\omega\sin\phi\,v - 2\omega\cos\phi\cos\theta w)\rho\Delta x\Delta y\Delta z$, where θ is the angle between the positive x axis and east, ϕ is the latitude and ω the angular velocity of the earth's rotation. The *Coriolis force* is an apparent force that is introduced into the equation of motion to allow a frame of reference to be used that is fixed relative to the rotating earth. The component due to the vertical velocity is maximum when the x axis lies along the east–west direction and zero when directed along a north–south line.

Now, Newton's second law of motion gives us: mass × acceleration = surface forces + body forces. The surface forces are the frictional forces on the boundaries, and the body forces are those affecting the whole water mass. Thus

$$\rho\Delta x\Delta y\Delta z\left(\frac{\partial u}{\partial t} + u\frac{\partial u}{\partial x} + v\frac{\partial u}{\partial y} + w\frac{\partial w}{\partial z}\right) = -\frac{\partial p}{\partial x}\Delta x\Delta y\Delta z$$

$$+ (2\omega\sin\phi\,v - 2\omega\cos\phi\cos\theta w)\rho\Delta x\Delta y\Delta z$$

$$+ \mu\Delta x\Delta y\Delta z\left(\frac{\partial^2 u}{\partial t^2} + u\frac{\partial^2 u}{\partial x^2} + v\frac{\partial^2 u}{\partial y^2} + w\frac{\partial^2 w}{\partial z^2}\right)$$

Therefore the instantaneous longitudinal equation of motion is

$$\frac{\partial u}{\partial t} + u\frac{\partial u}{\partial x} + v\frac{\partial u}{\partial y} + w\frac{\partial u}{\partial z} = -\frac{1}{\rho}\frac{\partial p}{\partial x} + f_1 v - f_2 w\cos\theta + \frac{\mu}{\rho}\left(\frac{\partial^2 u}{\partial x^2} + \frac{\partial^2 u}{\partial y^2} + \frac{\partial^2 u}{\partial z^2}\right) \tag{7.1}$$

Similarly, the vertical dynamic balance is

$$\frac{\partial w}{\partial t} + u\frac{\partial w}{\partial x} + v\frac{\partial w}{\partial y} + w\frac{\partial w}{\partial z} = -\frac{1}{\rho}\frac{p}{\partial z} + g - f_2(u\cos\theta - v\sin\theta)$$

$$+ \frac{\mu}{\rho}\left(\frac{\partial^2 w}{\partial x^2} + \frac{\partial^2 w}{\partial y^2} + \frac{\partial^2 w}{\partial z^2}\right) \tag{7.2}$$

where $(u\cos\theta - v\sin\theta)$ is the east–west component of the horizontal flows and g the gravitational acceleration.

The lateral balance is

$$\frac{\partial v}{\partial t} + u\frac{\partial v}{\partial x} + v\frac{\partial v}{\partial y} + w\frac{\partial v}{\partial z} = -\frac{1}{\rho}\frac{\partial p}{\partial y} + f_1 u - f_2 w\sin\theta + \frac{\mu}{\rho}\left(\frac{\partial^2 v}{\partial x^2} + \frac{\partial^2 v}{\partial y^2} + \frac{\partial^2 v}{\partial z^2}\right) \tag{7.3}$$

It is not normally possible to apply these equations other than in a time-averaged situation. However, they are the basis for numerical modelling. Let us consider the tidal average.

Vertical Dynamic Balance

In Equation (7.2) the vertical Coriolis effect is negligible compared with the gravitational acceleration and turbulent fluctuations can be averaged out by taking time means. The turbulent stresses and the vertical acceleration terms are also normally assumed to be negligible. Thus Equation (7.2) becomes the *hydrostatic equation*

$$-\frac{1}{\rho}\frac{\overline{\partial p}}{\partial z} = g \qquad (7.4)$$

The hydrostatic assumption is universally made both in the open ocean and in estuaries, except where finite amplitude internal wave motions are present.

Longitudinal Dynamic Balance

The horizontal, lateral and vertical velocities can each be considered as the sum of three constituents: a time mean over a tidal cycle, a simple harmonic velocity fluctuation of tidal period and a short period turbulent fluctuation. Substitution in Equation (7.1) gives

$$\frac{\partial \overline{u}}{\partial t} + \overline{u}\,\frac{\partial \overline{u}}{\partial x} + \overline{v}\,\frac{\partial \overline{u}}{\partial y} + \overline{w}\,\frac{\partial \overline{u}}{\partial z} + \frac{\partial}{\partial x}(\overline{UU}) + \frac{\partial}{\partial y}(\overline{UV}) + \frac{\partial}{\partial z}(\overline{VW}) =$$
$$-\left(\frac{1}{\rho}\frac{\partial p}{\partial x}\right) + f_1 v - \frac{\partial}{\partial x}(\overline{u'u'}) - \frac{\partial}{\partial y}(\overline{u'v'}) - \frac{\partial}{\partial z}(\overline{u'w'}) \qquad (7.5)$$

The molecular stress (viscosity) terms will be negligible compared with the turbulent stresses. Because the vertical velocities are small, the second Coriolis term can also be neglected. The various other cross-products have been neglected for reasons similar to those considered for the salt balance.

In Equation (7.5) the non-linear inertial terms associated with the tidal motion, such as $\partial/\partial x(\overline{UU})$, arise because the mean value of U^2 over a tide is not zero. If U is a simple harmonic function then $\overline{U^2} = U_0^2\,\overline{\cos^2 \omega t} = 0.5U_0^2$ and $\partial/\partial x(\overline{UU}) = U_0 \partial U_0/\partial x$. If the tidal wave is purely progressive then U and W will be 90° out of phase and the term $\partial/\partial z(\overline{UW})$ disappears. This term could also be neglected if the amplitudes of the tidal motions do not vary significantly with depth. This is unlikely, however, as the amplitude of both U and W will be reduced near the bottom. In a standing wave system the vertical and horizontal tidal velocities will be in phase and consequently $\overline{UW} = 0.5U_0 W_0$. Then

$$\frac{\partial}{\partial x}(\overline{UU}) + \frac{\partial}{\partial z}(\overline{UW}) = 0.5\left(U_0\,\frac{\partial U_0}{\partial x} - W_0\,\frac{\partial U_0}{\partial z}\right)$$

The term $\partial/\partial y(\overline{UV})$ disappears if lateral homogeneity is assumed.

The pressure term $\overline{((1/\rho)\partial p/\partial x)}$ is normally taken to be represented by $((1/\overline{\rho})\partial\overline{p}/\partial x)$, which assumes that pressure and density fluctuations in the water are uncorrelated.

The pressure at any depth is composed of two parts, one due to the density distribution and the other due to the slope of the free surface. The density distribution is obtained from salinity and temperature measurements, but the surface slope is normally obtained by calculation using a known or assumed level at which $\partial\overline{p}/\partial x$ is zero.

From the hydrostatic Equation (7.4), the pressure at any depth z is the atmospheric pressure plus the weight of water above that level

$$p = p_a + g \int_{-\zeta}^{z} \rho \, dz$$

where p_a is the atmospheric pressure, which can be assumed uniform, and ζ is the surface elevation above a datum level.

Thus the horizontal pressure gradient at any depth z is

$$\frac{\partial p}{\partial x} = g \int_{-\zeta}^{z} \frac{\partial \rho}{\partial x} \, dz - g\rho_s \frac{\partial \zeta}{\partial x} \tag{7.6}$$

where ρ_s is the density at the sea surface.

In a homogeneous sea the first term on the right of Equation (7.6) is zero and the pressure gradient is caused by the sea surface slope and its value is independent of depth. As a consequence, the surfaces of equal pressure are non-level throughout the depth. Where the density increases with depth the integral term also increases with depth so that the contribution of the sea surface slope is gradually compensated. Flow induced by the first, density, part of Equation (7.6) is called *baroclinic flow* and that due to the sea surface gradient is *barotropic flow*.

The integral term on the right-hand side of Equation (7.6) can be written in the form normally used in dynamical oceanography by replacing the density by specific volume $\alpha = 1/\rho$ and integrating with respect to p. This is to take account of the fact that the actual depth in the ocean is measured by the pressure.

Thus

$$\int_{-\zeta}^{z} \frac{\partial \rho}{\partial x} \, dz = -\frac{\langle \rho \rangle}{g} \frac{\partial}{\partial x} \int_{p_a}^{p_z} \alpha \, dp$$

where $\langle \rho \rangle$ is the mean density between the surface and the depth z, and Equation (7.6) becomes

$$\frac{1}{\rho}\frac{\partial p}{\partial x} = -\frac{\partial}{\partial x}\int_{p_a}^{p_z} \alpha\,\mathrm{d}p - g\frac{\partial \zeta}{\partial x} \qquad (7.7)$$

The integral on the right-hand side of Equation (7.7) is the expression for the dynamic depth of the pressure p_z

$$\boldsymbol{D} = \int_{p_a}^{p_z} \alpha\,\mathrm{d}p$$

The second term, the longitudinal surface water slope, can be measured by accurate surveying, but the tidally averaged slopes are likely to have considerable errors because they will be the differences between two large numbers. Alternatively, its value can be calculated if $\partial p/\partial x = 0$ at a depth $z = H$. This is done in the deep ocean, but in estuaries it is normally not possible and the surface water slope remains as a major unknown to be determined by other means.

The horizontal gradient of the dynamic depth $\overline{\boldsymbol{D}}$ can be calculated from the distribution of tidal mean salinity and temperature with depth using standard oceanographic tables, such as those of Unesco (1987). Alternatively, the baroclinic term can be calculated from the density distributions.

The density can normally be assumed to be a linear function of salinity in an *equation of state* $\rho = \rho_0(1+\alpha s)$, where ρ_0 is the density of fresh water at the appropriate temperature. When the salinity s is in ‰, $\alpha \approx 7.8 \times 10^{-4}$. Thus density gradients are approximately 0.78 the salinity gradients. An additional term can be added when suspended sediment concentration is high, so that $\rho = \rho_0(1 + 7.8 \times 10^{-4}s + 6.2 \times 10^{-4}C)$, where C is in $\mathrm{kg\,m^{-3}}$. Concentrations have to be very high to have an appreciable effect on density.

The final three terms on the right-hand side of Equation (7.5) are the frictional forces due to turbulent effects, the Reynolds stresses. As discussed in Chapter 4, they are written in the form

$$\overline{\tau}_{xx} = \rho(\overline{u'u'}) = -\rho N_x\frac{\partial \overline{u}}{\partial x}$$

$$\overline{\tau}_{xz} = \rho(\overline{u'w'}) = -\rho N_z\frac{\partial \overline{u}}{\partial z} \qquad (7.8)$$

$$\overline{\tau}_{xy} = \rho(\overline{u'v'}) = -\rho N_y\frac{\partial \overline{u}}{\partial y}$$

The coefficients N_x, N_z and N_y are the coefficients of eddy viscosity in the respective directions. It is generally possible to neglect the term involving N_x. Also, by assuming lateral homogeneity, so that $\partial \overline{u}/\partial y$ is zero, the turbulent frictional terms in Equation (7.5) reduce to that in the vertical direction

$$-\frac{\partial}{\partial z}(\overline{u'w'}) = -\frac{1}{\rho}\frac{\partial \bar{\tau}_{xz}}{\partial z} = \frac{\partial}{\partial z}\left(N_z \frac{\partial \bar{u}}{\partial z}\right) \tag{7.9}$$

Consequently, there are two particular problems that have to be resolved in determining the dynamic balance. The first is the value of the vertical eddy viscosity and the second is the surface water slope. If one is known, then the other can be calculated as the only unknown in the equation, provided sufficient measurements are available to quantify the other terms.

The value of N_z with depth depends on the distribution of shear stress. For unstratified conditions a linear variation can be taken between the surface and bed stresses. At the surface and bottom the values of τ_{xz} will be given by the longitudinal component of the net flux of momentum across the boundaries. At the surface it will be equal to the component of the mean wind stress and at the bottom it will be equal to the mean bed shear stress ($\bar{\tau}_0$). The bed stress can be determined in terms of a frictional coefficient k times the square of the depth mean velocity, i.e. *quadratic friction*. Thus $\tau_0 = k\langle u_z\rangle|\langle u_z\rangle|$. The modulus form $|\langle u_z\rangle|$ is required to preserve the sign of the oscillating current. Values of k are normally in the range 0.002–0.005.

Alternatively, the bottom stress can be calculated from the velocity profile using the von Karman–Prandtl formula (Equation 4.7)

$$u_Z = \frac{1}{\kappa}\left(\frac{\tau_0}{\rho}\right)^{1/2}\ln\frac{z}{z_0}$$

where z_0 is the bottom roughness length (the height above the bed at which the velocity is zero) and κ is the von Karman constant $= 0.4$. The velocity u_z is measured at a height z above the bottom. Though derivation of this formula assumes a steady flow, constant shear stress above the bottom and neutral conditions, profiles fitting the formula are commonly observed within 2 m or so of the bed. For a linear shear stress variation through depth the value of $N_z = N_0 = \kappa u_* z$ in a layer near the bed, so that N_z reaches a maximum at near mid-depth and then decreases to zero at the surface.

Though the shearing stress will be linear from the surface to the bottom in a steady current of uniform density, in a time varying flow no direct inference can be drawn about the distribution of the shearing stress with depth. However, with good current velocity measurements the profiles of shear stress can be calculated. In this context the equation of motion for a depth z becomes

$$\frac{\partial u_z}{\partial t} = -g\frac{\partial \zeta}{\partial x} + \frac{1}{\rho}\frac{\partial \tau_{xz}}{\partial z} \tag{7.10}$$

Taking depth mean values and assuming that the wind stress is zero

$$\frac{\partial \langle u\rangle}{\partial t} = -g\frac{\partial \zeta}{\partial x} - \frac{\tau_0}{\rho h} \tag{7.11}$$

The effect of the surface water slope can be removed by considering the difference between the depth-averaged equation and that at individual depths. The difference between Equations (7.10) and (7.11) gives

$$\frac{\partial}{\partial t}(u_z - \langle u \rangle) = \frac{1}{\rho}\frac{\partial \tau_{xz}}{\partial z} + \frac{\tau_0}{\rho h}$$

For finite time increments Δ_t

$$\frac{1}{\rho}\frac{\partial \tau_{xz}}{\partial z} = -\frac{\tau_0}{\rho h} + A_x$$

where the acceleration

$$A_x = \frac{\Delta_t(u_z - \langle u \rangle)}{\Delta_t} \tag{7.12}$$

Integrating (7.12) with respect to z gives:

$$\tau_{xz} = -\tau_0 \frac{z}{h} + \rho \int_0^z A_x \, dz$$

Thus the distribution of τ_{xz} with depth can be calculated from a knowledge of the bottom stress and the acceleration term A_x at various depths.

In Red Wharf Bay, Bowden *et al.* (1959) applied the above treatment to half-hourly data. They found the bed shear stress varied cyclically with maximum values of about $0.8\,\mathrm{N\,m^{-2}}$ ($8\,\mathrm{dynes\,cm^{-2}}$). The curves of τ_{xz} as a function of depth showed an almost linear increase from the surface to the bottom at times near maximum flood and ebb when the acceleration terms were small. When the flow was accelerating and the stress increasing, the stresses at mid-depths were less than those corresponding to a linear variation and the curve of stress against depth was concave upwards. When the current was decelerating this effect was reversed: the stress at intermediate depths was greater than a linear variation and the curves were convex upwards.

The values of the coefficients of eddy viscosity were also calculated. The highest values occurred at mid-depth when the current was greatest, with a maximum value of about $500\,\mathrm{cm^2\,s^{-1}}$. On the flood tide the mean value of N_z was about $270\,\mathrm{cm^2\,s^{-1}}$ and on the ebb about $130\,\mathrm{cm^2\,s^{-1}}$. On dimensional grounds the maximum value of N_z would be proportional to the tidal velocity times the depth. In this case the average value at mid-depth of $N_z = N_0 = 2.5 \times 10^{-3} U_0 h$. However, these short-term eddy coefficients cannot be directly compared with those obtained from results averaged over a tidal period, which have a slightly different physical meaning.

In neutral conditions $K_z \approx N_z = N_0$, but when a density stratification is present the vertical turbulence is inhibited and both K_z and N_z will be reduced. As the rate of generation of turbulent energy must exceed the rate of increase of potential energy due to vertical mixing K_z will be reduced more than N_Z. Consequently, the influence of stratification can be approached by using the theoretical relationships between the eddy coefficients and the Richardson number discussed in Chapter 4 [e.g. Equation (4.14)].

The effect of density stratification can be quantified in the same way as for neutrally stratified conditions by considering differences from depth mean values.

Assuming that lateral velocities are small and that $f_1 v$ is negligible, the equation of motion in the longitudinal sense at any depth can be written as

$$\frac{\partial u_z}{\partial t} = -g \frac{\partial \zeta}{\partial x} - gP - \frac{1}{\rho} \frac{\partial \tau_{xz}}{\partial z} \tag{7.13}$$

where

$$P = \frac{1}{\rho} \int_0^z \frac{\partial \rho}{\partial x} \, dz$$

Integrating from $z = 0$ to $z = h$ and assuming the surface stress to be zero

$$\frac{\partial \langle u \rangle}{\partial t} = -g \frac{\partial \zeta}{\partial x} - g \langle P \rangle - \frac{\tau_{xz}}{\rho h} \tag{7.14}$$

Taking differences between Equations (7.13) and (7.14)

$$\frac{\partial (u_z - \langle u \rangle)}{\partial t} = -g(P - \langle P \rangle) - \frac{1}{\rho} \frac{\partial \tau_{xz}}{\partial z} + \frac{\tau_0}{\rho h} \tag{7.15}$$

and

$$\frac{\partial (u_z - \langle u \rangle)}{\partial t} = \frac{\partial u_V}{\partial t} \quad \text{as} \quad u_z = \langle u \rangle + u_v$$

An equation similar to (7.15) can be developed which applies to the mean values over a tidal cycle. In the steady state if $\partial \bar{u}_V / \partial t = 0$, then

$$\bar{\tau}_{xz} = -g \int_0^z \rho(\bar{P} - \langle P \rangle) \, dz + \frac{\bar{\tau} z}{h} = -\rho N_z \frac{\partial \bar{u}_V}{\partial z} \tag{7.16}$$

Bowden (1960) has applied this approach to data obtained at a station in the centre of the Mersey Narrows during two periods. The values of $(\bar{P} - \langle P \rangle)$ were determined from the density distribution and the shearing stress $\bar{\tau}_{xz}$ on the first occasion was taken as zero when $\partial \bar{u}_1 / \partial z = 0$. This occurred at a depth z/h of 0.75. On the second occasion $\bar{\tau}_{xz}$ was taken as zero at the bottom. The values calculated for $\bar{\tau}_{xz}$ and N_z given in Table 7.1 are values related to the tidal cycle means and could be termed the effective shear stress and effective vertical eddy viscosity. Maximum values occur at mid-depth, but are only about one-tenth

Table 7.1 Values of effective shearing stress, eddy viscosity and eddy diffusion coefficients for two periods in the Mersey Narrows.

	1st period			2nd period		
z/h	$\overline{\tau}_{xz}$ $(\mathrm{N\,m^{-2}})$	N_z $(\mathrm{cm^2\,s^{-1}})$	K_z $(\mathrm{cm^2\,s^{-1}})$	$\overline{\tau}_{xz}$ $(\mathrm{N\,m^{-2}})$	N_z $(\mathrm{cm^2\,s^{-1}})$	K_z $(\mathrm{cm^2\,s^{-1}})$
0.1	0.02	9	5	0.034	14	8
0.3	0.045	27	11	0.086	46	23
0.5	0.039	40	27	0.103	73	30
0.7	0.010	43	17	0.086	72	29
0.9	−0.042	62	3	0.034	25	13

Reproduced with permission from Bowden, K.F. *I.A.S.H. Comm. Surface Waters*, Publ. 51, 1960, table 3.

of those which would be expected in a tidal current of uniform density. Values of the effective eddy diffusion coefficient K_z were calculated for the same data (Table 7.1) using Equation (6.23) and are about half the corresponding N_z. During the tidal cycle, however, instantaneous values for N_z and for K_z are likely to be much larger than these values.

Further data from the Mersey has been treated in the same way by Bowden and Sharaf el Din (1966a). In this case the Coriolis term was retained and the value of $\overline{\tau}_0$ was estimated on the assumption that $\overline{\tau}_{xz}$ was zero when $\partial \overline{u}/\partial z = 0$. This occurred at a depth $z/h = 0.8$. The calculated values of the effective shearing stress and the effective coefficient of vertical eddy viscosity are given in Table 7.2. The values of N_z tended to reach a maximum at mid-depth, but more consistent results were obtained by using cross-sectional mean values in Equation (7.16). These values of N_z are greater than the corresponding values of the effective vertical eddy diffusion coefficient calculated using Equation (6.23).

Pritchard (1956) examined the dynamic balance in the James River. He considered steady-state conditions and zero lateral velocities. It was argued from the salt balance analysis that the horizontal eddy flux of salt was small and Pritchard assumes by analogy that the horizontal eddy flux of momentum was also negligible. Consequently, Equation (7.5) becomes

$$\overline{u}\,\frac{\partial \overline{u}}{\partial x} + \overline{w}\,\frac{\partial \overline{u}}{\partial z} + U_0\,\frac{\partial U_0}{\partial x} = -\frac{\partial \overline{D}}{\partial x} - g\,\frac{\partial \zeta}{\partial x} - \frac{\partial}{\partial z}\,(\overline{u'w'}) \tag{7.17}$$

From the velocity data the acceleration terms on the left-hand side were calculated and from the salinity and temperature distribution also the relative pressure forces. To solve the equation for $\overline{\tau}_{xz}$, two boundary conditions are required to resolve the surface water slope and the Reynolds stress term. Those used were the surface and bed stresses. He evaluated the absolute pressure field

Table 7.2 Effective $\bar{\tau}_{xz}$, N_z and K_z for the Mersey Narrows.

z/h	Station 1			Station 2			Station 3			Mean across section	
	$\bar{\tau}_{xz}$ (N m^{-2})	N_z (cm^2 s^{-1})	K_z (cm^2 s^{-1})	$\bar{\tau}_{xz}$ (N m^{-2})	N_z (cm^2 s^{-1})	K_z (cm^2 s^{-1})	$\bar{\tau}_{xz}$ (N m^{-2})	N_z (cm^2 s^{-1})	K_z (cm^2 s^{-1})	$\bar{\tau}_{xz}$ (N m^{-2})	N_z (cm^2 s^{-1})
0.1	0.003	1*	8	0.031	240*	6	0.019	8	5	0.014	8.5
0.3	0.007	15	12	0.069	4	15	0.042	14	13	0.033	20
0.5	0.011	8	22	0.060	23	28	0.037	14	17	0.035	27
0.7	0.009	23	22	0.028	113	23	0.016	11	9	0.019	29
0.9	-0.011	5	1	-0.030	49	3	-0.016	22	1	-0.021	20

Reproduced by permission of Blackwell Science from Bowden, K.F. and Sharaf el Din, S.I. *Geophys. J. R. Astron. Soc.*, **10**, 1966, tables 3 and 4.
*Figures are uncertain and omitted from mean.

distribution and the level pressure surface existed near mid-depth. Above this depth the pressure surfaces sloped towards the sea and below it they sloped towards the head of the estuary. The longitudinal level pressure surfaces were above or below the depth of no mean motion depending mainly on the sign and the magnitude of the term $U_0 \partial U_0 / \partial x$. When this term was small the level pressure surface was nearly the same as the depth of no mean motion. If the tidal velocity amplitude U_0 increased downstream, then the depth would be greater than the depth of the no mean motion and vice versa.

The longitudinal component of the pressure force was mainly balanced by the eddy frictional term. As the Reynolds stress term is obtained by summing the other terms in Equation (7.5) it includes accumulated errors and the effects of terms not otherwise considered. However, a significant portion of the pressure force was balanced by the field change in the amplitude of the tidal velocity. This contrasts with the situation in fjords. The various longitudinal acceleration terms appear to be insignificant, also contrasting with the results in the surface layer of fjords.

Lateral Dynamic Balance

For average values over a tidal cycle the lateral dynamic balance will be

$$\frac{\partial \bar{v}}{\partial t} + u \frac{\partial \bar{v}}{\partial x} + \bar{v} \frac{\partial \bar{v}}{\partial y} + \bar{w} \frac{\partial \bar{v}}{\partial z} + \frac{\partial}{\partial x} \overline{(VU)} + \frac{\partial}{\partial y} \overline{(VV)} + \frac{\partial}{\partial z} \overline{(VW)}$$
$$= - \left(\overline{\frac{1}{\rho} \frac{\partial p}{\partial y}} \right) + f_1 \bar{u} - f_2 \bar{w} \sin\theta - \frac{\partial}{\partial x} \overline{(v'u')} - \frac{\partial}{\partial y} \overline{(v'v')} - \frac{\partial}{\partial z} \overline{(v'w')}$$

$$(7.18)$$

In any estuary the lateral and longitudinal tidal fluctuations are likely to be in phase so that in a purely progressive tidal wave situation the tidal terms will be

$$0.5 \left(U_0 \frac{\partial V_0}{\partial x} + V_0 \frac{\partial V_0}{\partial y} \right)$$

For a standing wave they will be

$$0.5 \left(U_0 \frac{\partial V_0}{\partial x} + V_0 \frac{\partial V_0}{\partial y} - W_0 \frac{\overline{\partial V_0}}{\partial z} \right)$$

However, these terms are generally not considered. The vertical Coriolis term is also considered negligible. The Reynolds stresses in this case are written

$$\bar{\tau}_{yx} = \rho\,(\overline{v'u'}) = -\rho N_x\,\frac{\partial \bar{v}}{\partial x}$$

$$\bar{\tau}_{yy} = \rho\,(\overline{v'v'}) = -\rho N_y\,\frac{\partial \bar{v}}{\partial y}$$

$$\bar{\tau}_{yz} = \rho(\overline{v'w'}) = -\rho N_z\,\frac{\partial \bar{v}}{\partial z}$$

The dominant term of the three stresses is likely to be that involving τ_{yz}. The pressure term can be again related to a depth of no motion, but it may not necessarily be the same as that in the horizontal sense, as the lateral sea surface slope will be different from that in the longitudinal sense. It is also necessary to include the effects of centrifugal force, as discussed in Chapter 5. Consequently, for steady-state conditions

$$u\,\frac{\partial \bar{v}}{\partial x} + \bar{v}\,\frac{\partial \bar{v}}{\partial y} + \overline{w}\,\frac{\partial \bar{v}}{\partial z} = -\frac{\partial \overline{D}}{\partial y} - g\,\frac{\partial \zeta}{\partial y} - f_1\bar{u} - \frac{1}{\rho}\,\frac{\partial \bar{\tau}_{yz}}{\partial z} + \frac{\bar{u}^2 + \overline{U^2}}{r_{xy}} \qquad (7.19)$$

where r_{xy} is the radius of curvature of the streamlines of the longitudinal flow in the y direction. The terms on the left are those that relate to the secondary circulation and they result from an imbalance in the terms on the right-hand side.

To see whether curvature of streamlines consistent with the estuarine topography is important in the balance between the various forces, a thin near-surface layer can be considered. If the gradient of the shearing stresses is taken as constant with depth, then the terms involving the sea surface slope, which is also constant with depth, and the Reynolds stresses can be eliminated by taking the differences between each term calculated at two separate depths. Thus Equation (7.19) can be written

$$\Delta_z\!\left(\bar{u}\,\frac{\partial \bar{v}}{\partial x}\right) + \Delta_z\!\left(\bar{v}\,\frac{\partial \bar{v}}{\partial y}\right) + \Delta_z\!\left(\overline{w}\,\frac{\partial \bar{v}}{\partial z}\right) = \Delta_z\!\left(\frac{\partial \overline{D}}{\partial y}\right) - \Delta_z(f_1\bar{u}) + \Delta_z(\bar{u}^2 + \overline{U^2}/r)$$

$$(7.20)$$

where Δz is the operator $\partial/\partial z$. The assumption of a linear variation of shearing stress with depth is reasonable in the near-surface layer where conditions are fairly uniform. The radius of curvature of the streamlines is then the only unknown. Once this is calculated, the surface water slope can be calculated assuming that the stress is negligible.

The above approach has been applied to the surface layer of the Vellar estuary and Southampton Water by Dyer (1977). This showed that the dominant terms in the balance are the water slope, the internal density structure and the centrifugal force. The Coriolis force was of secondary importance. For larger estuaries, however, the Coriolis force is likely to become relatively more important as the radius of curvature of the estuary bends become larger. In the entrance to the Hudson River estuary, which is over 90 km wide, Doyle and Wilson (1978) found that the lateral pressure gradient balanced the sum of the centrifugal and Coriolis forces. However, over much of the transect Coriolis force was of secondary importance.

In salt wedge estuaries the internal density field will be negligible and the lateral balance in the surface layer will be between the surface water slope and the centrifugal force, similar to rivers, as we have seen in Chapter 5.

In fjords analysis is often made by assuming a lateral balance between Coriolis force and the pressure gradients

$$f_1 \bar{u} = -\left(\frac{1}{\rho} \frac{\partial \bar{p}}{\partial y} \right) \tag{7.21}$$

Thus the lateral accelerations, the frictional stresses and the body forces are considered negligible. Providing some assumption can be made about the depth of no motion, the depth of a surface with zero horizontal pressure gradients, Equation (7.21) can be solved. Cameron (1951), from observations at the entrance to Portland Inlet, British Columbia, used a depth of no motion of 27.4 m and obtained satisfactory agreement between the calculated and observed fresh water discharge.

This process appears to be valid in other straight, deep fjords. Tully (1958), with the same basic assumptions, examined the transports in the Juan de Fuca Strait. The fresh water fraction f of the total seaward volume transport Q_1 in the upper zone equals the river flow

$$f Q_1 = R$$

The landward transport Q_2 of sea water in the lower layer

$$Q_2 = (1 - f)Q_1 = \frac{1 - f}{f} R$$

The lateral balance gives

$$\frac{\partial \bar{D}}{\partial y} = f_1 (\bar{u}_1 - \bar{u}_2)$$

As

$$\bar{u}_1 - \bar{u}_2 = \frac{Q_1}{A_1} - \frac{Q_2}{A_2}$$

Then

$$Q_1 = \frac{\partial \bar{D}}{f_1 \partial y} \frac{A_1 A_2}{A_2 + A_1(1 - f)} \quad \text{and} \quad Q_2 = -\frac{\partial \bar{D}}{f_1 \partial y} \frac{A_1 A_2 (1 - f)}{A_2 + A_1(1 - f)} \tag{7.22}$$

In this case the depth of no motion, to which the dynamic height anomalies were computed, and the depth for separation of the areas A_1 and A_2 were obtained from the salinity distribution. A plot of salinity against logarithmic depth gave three straight lines: the upper and lower almost isohaline zones were separated by a halocline. The junction between the halocline and the lower layer was remarkably constant and it was argued that this must be a level where transport only takes place vertically. The dynamic depth anomalies were consequently related to this depth. The salinity at this depth was called the index salinity and was the base salinity used for calculation of the fresh water fraction in the upper layer.

This method of analysis could be used indirectly to measure the tidal velocities, as Tully also pointed out that in many cases the velocity gradient remains nearly constant throughout the tidal cycle. The transverse slope of the pressure surfaces must then be almost constant and instantaneous values as well as mean values could be used in Equation (7.21).

Solution of the general equations of motion in estuaries is more difficult than for the equation of salt continuity. In the latter case we were dealing with the product of a vector and a scalar, whereas with the former we have the product of two vectors. As a consequence we have to be more accurate with our direction measurements and more stringent about our averaging criteria. To diminish the effects of errors in the current direction what is really required is that the x coordinate should be directed along the current flow, rather than along the estuary, and, as we have seen, the two may be locally rather different. There can be drastic variations of flow direction during a tide, the flood tide need not be 180° different in direction from the ebb tide. This effect can also alter with depth and will be especially apparent with the residual flows, as the control on the surface mean flow is mainly the upstream topography and on the bottom flow the topography downstream. Therefore, unless careful measurements are made, directional errors can produce rather large errors in the lateral velocities and these will be exaggerated in the calculation of the vertical velocities. Because of the importance of topographic controls on the flow in the lateral direction, estimation of the turbulent stresses is virtually impossible except by direct measurement. Accurate measurements are needed to validate mathematical models and it is apparent from the above analysis that a good representation of the topography is necessary. The choice of eddy coefficients has to be matched to the number of dimensions of the model and it must be remembered that the eddy coefficients are dispersion coefficients and include the effects of spatial averaging.

It is of interest to compare the depth mean values of the eddy coefficients obtained from different estuaries with the different estuarine characteristics. Table 7.3 shows the results from six estuaries ranging from well mixed to highly stratified, as shown by the surface to bed salinity difference δs. The eddy viscosity and S_t appears to scale inversely with stratification, though relatively high values occur in the Mersey and the Columbia River estuaries because of

Table 7.3a Observed values

Location	δs	u_s (cm s^{-1})	u_b (cm s^{-1})	$\partial s/\partial x$ (km^{-1})	h (m)	$\dfrac{S}{F}$	$\dfrac{\langle u \rangle}{U_0}$	S_t
Long Island Sound	0.8	22	-5	0.035	31	—	—	—
Mersey Narrows	0.9	12	-9	0.238	20.4	83	0.001	92860
Southampton Water	1.46	8	-6	0.222	12	100	0.003	760
James estuary	3.0	12	-10	0.587	7	90	0.014	47
Columbia River estuary	15.0	29	-27	1.411	10.6	40	0.06	6
Vellar estuary	22.5	9	-4	2.014	2.7	0.1	2.0	3

Table 7.3b Computed values

Location	N_z (cm^2 s^{-1})	K_z (cm^2 s^{-1})	k ($\times 10^{-3}$)	$\mathrm{d}\zeta/\mathrm{d}x$ (cm km^{-1})	K_z/N_z
Long Island Sound	8.1	9.5	11	0.03	1.2
Mersey Narrows	41	25	4.5	0.16	0.6
Southampton Water	12	3.3	3.3	0.12	0.3
James estuary	4.4	2.2	0.8	0.14	0.5
Columbia River estuary	17	8	0.7	0.54	0.5
Vellar estuary	0.9	0.1	6.0	0.17	0.1

Reproduced by permission of the National Academy of Sciences from Officer (1977) and from Prandle (1985) by permission of Academic Press.

high tidal velocities. The eddy diffusivity shows similar trends. However, even in comparatively well-mixed estuaries the coefficients are smaller, often by an order of magnitude, than would be expected for tidal currents of that amplitude in homogeneous water. The ratio of the two, the Prandtl number, decreases with increasing stratification, as would be expected. From these data it is not possible to calculate meaningful Richardson numbers.

Chapter 8

Highly Stratified Estuaries

Highly stratified estuaries can be considered in three categories: arrested salt wedge, highly stratified with tidal motion and fjords. Some aspects of these have already been considered.

The classical view of salt wedge estuaries is that of the *arrested salt wedge*. In this the salt wedge is stationary and there is a balance in the upper layer between the pressure forces and the frictional forces. The pressure field is produced by the slopes of the water surface and the interface, and balances the interfacial friction. This obviously applies to the non-tidal or tidally averaged situation, where the tidal motion advects the structure within the estuary without appreciable change.

In some estuaries the river discharge is high enough to maintain high stratification even when the tidal motion is large enough to advect the wedge out of the estuary near to low water. In others the tidal advection may be resisted by the isolation of the saline water by sills in deeper basins. Nevertheless, the salt wedge is renewed by landward inflow during the flood and the salt water recedes and is lost by entrainment into the upper layer on the ebb. Thus at any point in the estuary there is a rise and fall of the density interface at a tidal frequency, though there may be considerable asymmetry in its motion, despite the fact that the tide is small relative to the influence of river discharge. This baroclinic tidal motion is a phenomenon that is not found to the same extent in weakly stratified and partially mixed estuaries, and may be a fundamental difference between highly and less well stratified estuaries (Jay and Smith 1990a).

In fjords the stratified flow at the mouth, or over shallow sills, will control the intrusion of salt water, but within the main part of the fjord the same dynamic balance as the arrested wedge is likely.

ARRESTED SALT WEDGE

Because of the entrainment both salt and volume have been added to the upper layer in its passage down the estuary. Consequently, its discharge increases towards the mouth. The slope of the density interface at the top of the salt wedge will be determined by the magnitude of the frictional stresses on it. These stresses vary with F_i and Re. From continuity considerations Stommel (1953a) has shown that the depth of the upper layer varies with the transport. Entrainment of salt water from below increases the thickness of the upper layer towards the mouth for F_i less than 0.5. For F_i greater than 0.5 and less than unity the thickness of the fresh surface layer decreases. A similar variation of upper layer thickness with changes in river discharge has been described by Tully (1949) in Alberni Inlet.

The length and shape of the salt wedge in an estuary has been examined by Farmer and Morgan (1953) and Ippen and Harleman (1961). Farmer and Morgan assumed an estuary without lateral variations, of regular depth and with a rectangular cross-section. They also assumed a sharp density discontinuity, no mixing between the layers and a negligible velocity in the salt wedge. They found that the shape of the wedge was represented by

$$kF_{i0}^2 \frac{x}{h_0} = \frac{1}{6} n^2(3 - 2n) - F_{i0}^2 \left(\frac{n}{1 - n} + \log(1 - n) \right) \qquad (8.1)$$

where h_0 and F_{i0} are the total water depth and the interfacial Froude number at the tip of the salt wedge, D is the thickness of the wedge at any position x, $n = D/h_0$, and k is the drag coefficient on the wedge. Measurement of D at the mouth, at which $x = L$, for varying values of the mean stream velocity at the wedge tip, enables the drag coefficient to be determined. There was agreement of the theory with laboratory experiments and with the south-west pass of the Mississippi for which a k value of 0.001 was determined. In experimental tests using a k value of 0.006 the predicted length of the wedge was within 15% of that observed. The wedge length increased with decreasing discharge and increased with increasing density contrast. A comparison between the predicted and observed wedge shapes is shown in Figure 8.1.

Ippen and Harleman (1961) and Harleman and Ippen (1960) examined the intrusion length in models of tidal estuaries. In partially mixed estuaries the intrusion length L was

$$L \propto \sqrt{\frac{g(\Delta\rho/\rho)}{u_f G^{1/3}}} \qquad (8.2)$$

where u_f is the fresh water flow rate and G is the rate of energy dissipation per unit mass of fluid in the estuary.

In well-mixed estuaries the intrusion length was

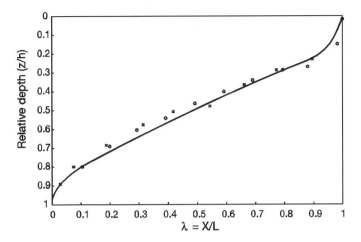

Figure 8.1 Profile of a salt wedge: comparison of results from theory and observations. *X*, Distance from the head of the wedge; *L*, length of wedge; solid line, theory; ○, Mississippi River Southwest Pass; and x, model experiment. Reproduced from Farmer and Morgan (1953). The salt wedge. In: *Proceedings of the 3rd Conference on Coastal Engineering* by permission of the American Society of Civil Engineers.

$$L \propto \frac{G^{1/3}}{u_f} \tag{8.3}$$

Rattray and Mitsuda (1974) carried out a theoretical analysis for a rectangular cross-section of constant depth and a sloping bottom. They assumed that there was a negligible salt flux between the layers and the horizontal gradients of density in each layer were small. In the upper layer the water must accelerate as it flows seawards over the lower layer as it is constrained laterally and decreases in thickness. The equation of motion is thus a balance between this acceleration, the barotropic pressure gradient and the friction at the interface. In the lower layer acceleration is unimportant and the motion is a balance between the barotropic and baroclinic gradients due to the longitudinal variation in layer thickness and the sea bed friction. A striking result is that for a wide range of Reynolds numbers the ratio of the bed shear stress to the interfacial stress changed only slightly as the ratio of the layer depths changed from a very small to a very large value. The wedge profile depended on the bottom topography and on the densimetric Froude number. As this is a function of the fresh water discharge, the density contrast between the layers, and the channel depth at the tip of the wedge, these three variables have to be known to a considerable degree of accuracy before reasonable predictions of the salt wedge length can be made. In fact, the entrainment rate is likely to increase along the wedge from the tip to the mouth.

More natural topographies have been considered by Prandle (1981). He modelled the salinity intrusion in estuaries with a breadth variation of $b(x/L)^n$, and a depth variation of $h(x/L)^m$, where L is the estuary length. The salinity distribution was highly dependent on the dimensionless parameter $u_1 x_1/K_x$, where K_x is a longitudinal dispersion coefficient and u_1 is the freshwater flow velocity at position x_1. Comparison with a number of estuaries showed that the salinity distribution was highly sensitive to the value for the dispersion coefficient, which was unlikely to be constant throughout the estuary, and that the variation of cross-sectional area is a fundamental parameter in determining the mixing pattern in an estuary.

Generic modelling of the salt intrusion therefore appears to have fundamental difficulties and specific examples need to be considered.

HIGHLY STRATIFIED WITH TIDAL MOTION

Despite the highly stratified nature of salt wedge estuaries there are considerable variations observed during a tidal cycle. Both upper and lower layers can oscillate, but the amplitude in the lower layer is small enough that the boundary turbulence is ineffective in mixing. Entrainment is the primary mixing process. The upper layer thickness will generally be at a maximum at about low water, but critical internal Froude numbers may be exceeded during the ebb tide, giving intense mixing.

We will now consider a few examples which will illustrate the processes that are active in creating the mixing and circulation which need to be incorporated in predictive tidal modelling.

Vellar Estuary, India

The Vellar estuary (11°29′ N lat., 79°47′ E long.) has been described by Dyer and Ramamoorthy (1969). The lowermost 6 km is straight, about 300 m wide, and averages about 1.5 m deep (Figure 8.2). Just inside the mouth lagoons exist on either side behind a wide sandy beach, the breakpoint bar forming the essential feature of a bar-built estuary. Within the estuary there is a series of alternating scour holes that are consistent with a meandering flow of water. The normal tide range in the estuary is about 0.7 m.

The distribution of salinity at high and low waters for three periods of decreasing river flow after a flood is shown in Figures 8.3, 8.4 and 8.5. During the flood tide there was an intrusion of a saline wedge along the bottom, bounded by a halocline of salinity 20 in 0.5 m, from the fresh river water on the surface. This wedge acted as a piston impounding river water in the upper estuary. At high water at the mouth the water column was only moderately stratified with a bottom salinity that increased during the three surveys. The distance of penetration of the salt wedge increased with time as the river

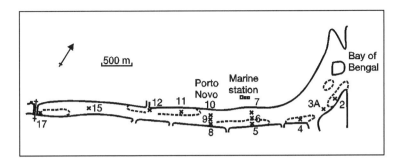

Figure 8.2 Vellar estuary showing location of stations. Dotted lines represent approximate limits of the deeper water areas. From Dyer and Ramamoorthy (1969). Reproduced with permission from the American Society of Limnology and Oceanography.

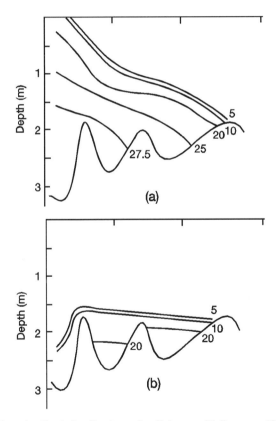

Figure 8.3 Longitudinal distribution of salinity on 20 January 1967, Vellar estuary, high river flow. (a) High water; (b) low water. From Dyer and Ramamoorthy (1969). Reproduced with permission from the American Society of Limnology and Oceanography.

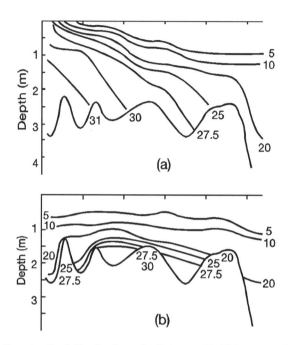

Figure 8.4 Longitudinal distribution of salinity on 27–28 January 1967, Vellar estuary, low river flow. (a) High water; (b) low water. From Dyer and Ramamoorthy (1969). Reproduced with permission from the American Society of Limnology and Oceanography.

discharge decreased from the initial flood stage. At about high water the impounded river water was released and quickly established a homogeneous surface layer with a virtually horizontal halocline throughout the estuary. This halocline was rapidly pushed downwards until it reached the level of the sills between the basins, the shallowest sill being 0.5 km inside the mouth (station 3A). The saline waters were then isolated within the basins, but under the influence of the fresh water flow the saline water was entrained from the top of the halocline, as well as flowing over the sills from one basin to another.

Certain of the basins seemed to be especially prone to flushing out by the fresh water on the ebb tide. It is apparent by comparison between the high and low tide bottom salinities that the most stable basins were those at stations 5, 6, 7 and station 17. The basins most prone to dilution of bottom salinity were those at stations 2 and 4.

Just before low water, as the downstream pressure gradient diminished, the seaward flowing surface water decreased in velocity and the saline water started moving landwards along the bottom at the mouth. It then passed upstream from one basin to the next, causing abrupt increases in bottom salinity. Development of this salt wedge pushed the halocline upwards and impounded the river flow in the upper estuary.

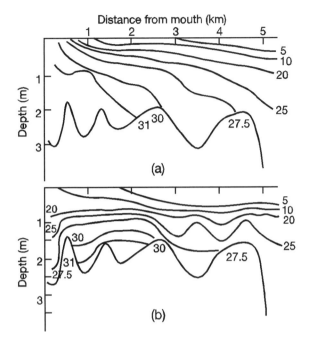

Figure 8.5 Longitudinal distribution of salinity on 9–10 February 1967, Vellar estuary, low river flow. (a) High water; (b) low water. From Dyer and Ramamoorthy (1969). Reproduced with permission from the American Society of Limnology and Oceanography.

The relationship between salinity and velocity variations during a tidal cycle is shown by the results for stations 2 and 7 on 17 January (Figure 8.6). At station 2 on the flood tide the current was at a maximum at mid-depth and the salinity increased regularly from the top to the bottom. About an hour before high water the surface current reversed and the salinity started decreasing. The bottom current, however, reversed at high water. During the ebb tide the maximum velocity was at the surface, but with measurable velocities at the bottom. About 0.75 hour before low water, the saline water started flowing in along the bottom, but the surface current continued to flow seaward until about 0.5 hour after low water. At station 7 the pattern was similar except that the current reversed throughout the column before high water and after low water. During the ebb the bottom velocities were immeasurably low and this coincided with the period of high bottom salinities caused by entrapment of the bottom water.

Typical curves of mean salinity and velocity calculated over a tidal cycle are shown for station 7 (Figure 8.7). The mean salinity at the first period (20 January) showed that a fresh surface layer overlay a thin salt wedge. This was associated with a seaward velocity throughout the water column. At this time

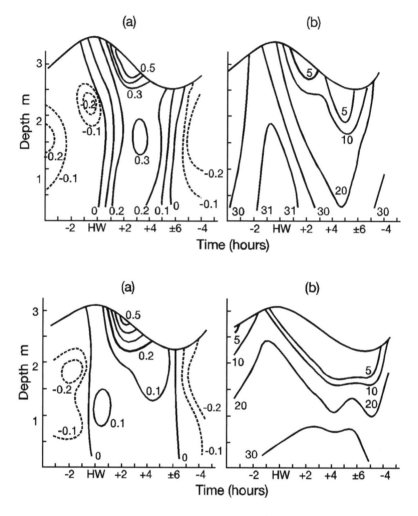

Figure 8.6 Variation of (a) longitudinal velocity (m s^{-1}) and (b) salinity on 27 January 1967. Upper panel, station 2; lower panel, station 7. From Dyer and Ramamoorthy (1969). Reproduced with permission from the American Society of Limnology and Oceanography.

the seaward velocity was greatest in the deeper part at station 7. In the shallower parts of the section the seaward mean velocities were less. Thus the isotachs were tilted downwards towards the right (Figure 5.1). As river discharge decreased, the saline bottom layer became thicker and more homogeneous and a landward mean velocity developed at mid-depth. At the last period the surface mean velocity had decreased abruptly and at station 7 was considerably less than the landward bottom flow. The necessary mean

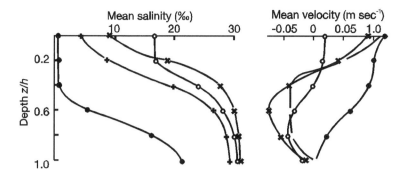

Figure 8.7   Tidal mean salinity and velocity profiles for station 7. (●) 20 January; (+) 27 January, (x) 9 February, and (○) 15 February. From Dyer and Ramamoorthy (1969). Reproduced with permission from the American Society of Limnology and Oceanography.

seaward flow due to river discharge was maintained by an increased mean surface flow and decreased mean landward bottom flow in the shallower waters of stations 5 and 6. Thus with decreased river discharge the depth of no motion and the isotachs became tilted downwards towards the left (see also Figure 5.1).

The values of the dimensionless stratification and circulation parameters were obtained for the section of stations 5–7 from the observed salinity and velocity values. The results are plotted in Figure 8.8 with the calculated river discharge rates in $m^3 s^{-1}$ in parentheses. This shows that at periods of high river discharge the estuary is a salt-wedge type. At decreased river flow the estuary becomes well stratified with the mean current reversing at depth. It is probable that as the river flow diminishes further the estuary would become less well stratified. In the hot season, with no river flow, the estuary may become homogeneous, though temperature may become important in producing density differences by surface heating.

Palmiet Estuary, South Africa

The characteristics of the tidal intrusion of the salt wedge in a small microtidal bar-built estuary, the Palmiet, South Africa has been described by Largier and Taljaard (1991). The measurements were carried out during a period of low river flow and high tidal range when the mouth of the estuary was open and there was tidal exchange through the mouth. The estuary comprises a single basin about 4 m deep and 200 m long. The tidal range reached about 1.5 m and tidal intrusion of salt seemed to be limited to river flows less than about $20\,m^3\,s^{-1}$.

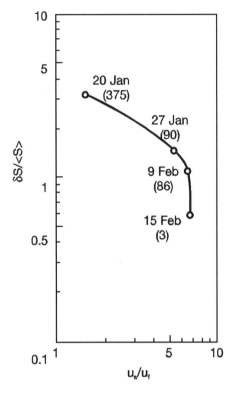

Figure 8.8 Classification of the Vellar estuary according to Hansen and Rattray (1966). River discharge rates in parentheses ($m^3 s^{-1}$). From Dyer and Ramamoorthy (1969). Reproduced with permission from the American Society of Limnology and Oceanography.

The sequence of events during the tide is shown in Figure 8.9. Immediately preceding the flood tide there was a layer of water in the basin of salinity 20, trapped from the previous high tide. At the beginning of the flood tide a tidal intrusion front occurred at the mouth and the denser saline water flowed as a density current down the slope of the basin. The level of the water surface was thus raised from below by the volume of the new sea water, as well as by the impounded river water. Thus both layers thickened. During the ebb tide the river water flowed out over the surface and gradually reduced the elevation of the interface of the trapped saline water by shear driven entrainment. This erosion had no effect on the salinity of the water close to the bed, being restricted to close to the halocline, as a consequence of the Richardson number being large. Thus as the river flow diminished, from tide to tide the salinity in the bottom layer and its thickness during the ebb tide increased. With an increase in river flow the tidal intrusion was prevented and all the saline water resident in the basin was removed.

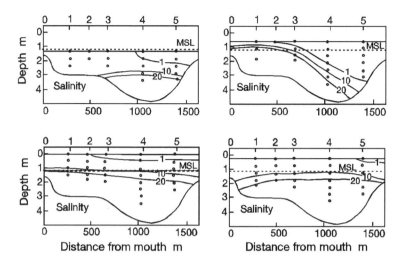

Figure 8.9 Longitudinal sections of salinity in the Palmiet estuary for low river flow. (a) Low tide; (b) flood tide; (c) high tide; and (d) ebb tide. Reproduced from Largier and Taljaard (1991) by permission of Academic Press.

On one day the sequence of events at the mouth is of interest. The barotropic flood tide started about three hours after low water, but lasted only for 75 minutes. However, before the depth-integrated tide turned, sea water intruded beneath the ebbing surface layer, with a two-way exchange flow carrying bottom water into the estuary. The intensity of the shear at the interface varied with time-scales of order of a minute, in response to fluctuations in flow in the surf zone outside the mouth. As the flood strengthened, a foam line and frontal structure occurred where the inflowing saline water arrested the outflowing surface water, and gradually pushed it from the seaward end of the entrance to the landward end, forming a V-shaped tidal intrusion front. The density current advanced within the estuary at a speed of $0.2–0.3 \, \text{m s}^{-1}$, which was reduced by the 'head wind' of the opposing seaward moving surface layer to about $0.17–0.26 \, \text{m s}^{-1}$. On plunging, the basal flow was supercritical and the vertical exchange in the hydraulic jump created a dilution of about $3:1$ in the underflow, reducing its salinity by about a third. Minimum dilution occurred at peak flood when the plunge line was furthest into the estuary. A critical R_i value was also recorded at the head of the gravity current, but the mixing products were contained in the stratified layer behind the head of the current.

During the ebb tide the halocline was characterized by a high R_i, and entrainment sharpened the surface layer of the halocline, not penetrating into the bottom layer, which maintained a constant salinity. Entrainment velocities of $4.2–8.3 \times 10^{-5}$ m s^{-1} were observed on the flood tide. On the ebb tide mean entrainment velocity was 3.5×10^{-5} m s^{-1}.

The mouth of the estuary becomes closed at low river flow and the saline water becomes trapped. Slinger and Largier (1990) determined that the two-layer structure could persist for up to 2.5 months after closure, with only molecular diffusion of heat and salt occurring in the deeper layer.

Fraser River Estuary, British Columbia

A contrasting situation to the two preceding examples is provided by the Fraser River estuary in British Columbia (Geyer and Farmer, 1989). This has an average discharge of 3000 m^3 s^{-1} , as well as being mesotidal with a range of 2.5–4.0 m, with the result that the salt wedge is strongly influenced by tidal action. The estuary is 10–15 m deep, with a delta-like mouth constrained by long training walls.

At low water a sharp, nearly vertical front located near the mouth separates saline from almost fresh water. The front may indicate internal hydraulic control, with subcritical conditions within the estuary changing to supercritical in the plume. The pycnocline thickens downstream as a result of mixing of water from the adjacent mixed layers into the zone of high shear.

During the flood tide the saline wedge propagates into the estuary as a density current, with a strong gently sloping pycnocline extending through the estuary. Behind the head there is a zone several kilometres long of intermediate density water. At the end of the flood tide the salt wedge reaches a temporary arrested condition and the interface becomes sharp with only a slight landward slope.

During the strong ebb the system becomes supercritical and salt water is rapidly mixed, high salinity water is confined to a thin bottom layer and this is gradually eroded until the estuary is fresh.

The salt wedge advance was monitored with a high frequency echosounder. Figure 8.10a shows the results obtained from an anchored vessel as the salt wedge propagated past. The abrupt raised head has the appearance of a gravity current, with a broad gradient zone of intermediate salinity behind, and in front of the homogeneous salt wedge. The head was made up of low salinity water created by mixing in the shear instabilities present at the bed. The propagation speed was estimated as being 0.4 m s^{-1} faster than the depth mean inflow resulting from the tide minus the river flow.

Figure 8.10b shows a profile along the front. Strong horizontal gradients appear to be confined to about 200 m behind the head, after which a homogeneous lower layer appears. The pycnocline slope varies due to changes in the bottom slope. Behind the head large amplitude variations of pycnocline elevation are in phase with sand wave features on the sea bed. Further than 0.6 km behind the head the pycnocline appears to be out of phase with the bed

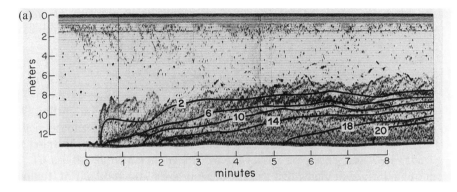

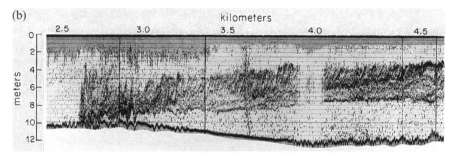

Figure 8.10 Echosounding traces across the advancing saline wedge in the Fraser River. (a) From a stationary vessel as the intrusion passes, with superimposed salinity contours and (b) from a moving vessel, showing Kelvin–Helmholtz waves near the head. Note that near the head the interface follows the bed features, indicating supercritical conditions. Reproduced from Geyer and Farmer (1989) by permission of the American Meteorological Society.

features. This suggests that the flow at the head of the wedge is supercritical and that of the main body of the wedge is subcritical.

The speed of advance of the salt wedge is similar to the depth-averaged velocity of the freshwater discharge, so that the trajectory of the salt wedge is consistent with the sum of the tidal and density-driven motions. When the river flow increases, its outflow velocity would therefore exceed that of the density current and the penetration of the salt wedge would be less than the tidal trajectory, and vice versa.

During the advance, Richardson number profiles show values close to 0.25 throughout the flow, except at the lower part of the pycnocline where the velocity profiles show a maximum (Figure 8.11). This 'jet' appears to result from the very reduced turbulence levels at the pycnocline and the low turbulent shear stresses, a low vertical exchange of momentum. This effect has been modelled by Geyer (1988). At the pycnocline the horizontal density-driven pressure gradient is balanced by the horizontal momentum. Nearer the bed

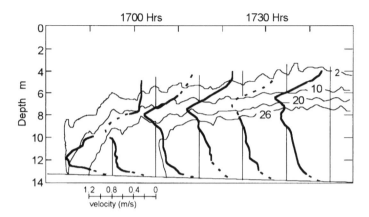

Figure 8.11 Velocity profiles superimposed on salinity contours during the advancing wedge in the Fraser River. Note that the velocity maximum is close to the centre of the pycnocline. Reproduced from Geyer and Farmer (1989) by permission of the American Meteorological Society.

friction is important in reducing the velocity. A consequence of this is that the brackish water at the pycnocline will tend to overtake the denser underlying water to accumulate near the head of the salt wedge. Thus the intermediate salinity zone behind the head of the wedge may be due to advection rather than being created by mixing.

During the ebb tide there are three distinct phases. The first is maintenance of the sharp, well-defined interface formed on the flood tide, with little vertical mixing. Though shear between the ebbing surface water and the almost stationary bottom water is large, it does not produce shear instability. The second phase is when the wedge is forced to retreat by the outward pressure gradient. Bottom friction is increased, leading to shear instabilities which thicken the pycnocline and produce a vertical flux of salt out of the bottom layer. Thus the wedge thins because of advection and vertical entrainment. The mixing tends to be localized to zones of shallower depth or to constrictions at which the flow becomes supercritical. However, no distinct hydraulic jumps were evident, perhaps because the Froude number was very close to critical throughout. The third phase was when the two-layer structure had diminished and only remained in deeper sections of the estuary. These were separated by mixed water and were eventually eroded if the ebb was strong enough.

INTENSE MIXING PERIODS

Where distinct hydraulic jumps occur on the ebb, *intense mixing periods* (IMP) can be produced. These were first described by Partch and Smith (1978) in the

Duwamish estuary in Washington State. During the flood and very early ebb there was a thin surface layer about 1 m thick, a strong pycnocline 1–2.5 m thick and a homogeneous bottom layer about 3 m thick. Figure 8.12 shows three tidal cycles of measurement at one station. There is considerable diurnal inequality so that the range of the second tide is only about half (1.5 m) of the other two. During the intense mixing events an abrupt increase in surface salinity occurred at the same time as a rapid descent of the interface. They result from an internal hydraulic jump caused by supercritical Froude number and this was calculated to occur over a reach of the estuary with about half of the salt wedge affected by the intense mixing. The central ebb tide did not produce an IMP because of lower currents and differing stratification. It was estimated that about 50% of the vertical salt transport occurred during the IMP and values of the vertical eddy diffusivity of salt in the pycnocline reached values of 5–6 cm^2 s^{-1}. After generation of an IMP the mixed water will be advected along the estuary.

Other measurements in the same estuary by Gardner and Smith (1978) defined the processes as being much more localized, with advection being important. About 66% of the mixing occurred in hydraulic jumps at steps in the bottom topography or at constrictions. Another 22% resulted from an internal hydraulic jump caused by the flow response to a bridge and obstruction to the flow caused by the bridge piers. The final 12% of mixing occurred in a 1 km reach of shallower water downstream of the bridge. Between the areas of supercritical flow the subcritical conditions may allow sharpening of the stratification by velocity shear without significant associated mixing.

Similar observations in the Tees estuary have also attributed an IMP to an internal hydraulic jump at a bridge, with subsequent advection of the structure along the estuary (New *et al.*, 1986, 1987). Consequently, mixing may be very localized in time and related to particular topographic features within the estuary.

FJORDS

Hardanger Fjord, Norway

The hydrography of Hardanger Fjord has been described by Saelen (1967). This fjord is over 100 km long and over 850 m deep and has a sill depth of about 150 m. River discharge averages 385 m^3 s^{-1} during June, but is less than 20 m^3 s^{-1} during February. The longitudinal distribution of temperature and salinity at high and low flows are shown in Figure 8.13.

In summer, with the high river discharge, a fresher surface layer is formed with a thickness of less than 20 m. At this time the surface water is warmer than that below. There is a possibility of an inflow of coastal water occurring at a depth of 10–100 m during the late summer, providing an additional heat supply

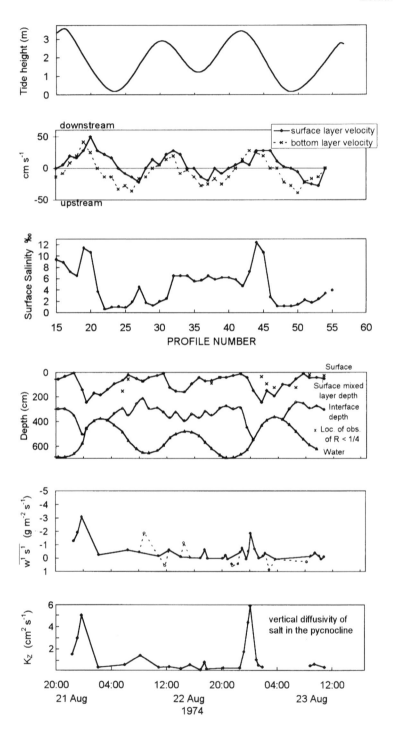

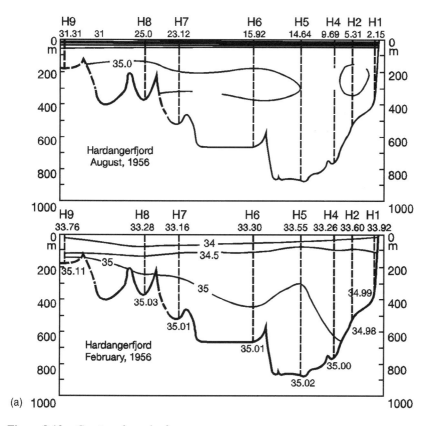

Figure 8.13 *Continued overleaf.*

to the surface layers. During the winter the surface salinity is only slightly lower than that at depth. The surface temperature is also low and there is a possibility that vertical thermohaline mixing may occur if the surface density is high.

In the winter the bottom temperatures are lower towards the mouth. In the summer the temperatures near the bottom at the mouth are below the winter

Figure 8.12 (*opposite*) Three tides in the Duwamish estuary. From the top: tide curves; tidal currents in the upper and lower layers; surface salinity, surface mixed layer depth, interface depth and water depth, with location of times of subcritical gradient *Ri*; vertical turbulent flux; and vertical eddy viscosity. Note the peaks of vertical turbulent flux and eddy diffusivity coincide with the interface thinning and the bottom layer disappearing. Reproduced from Partch and Smith (1978) by permission of Academic Press.

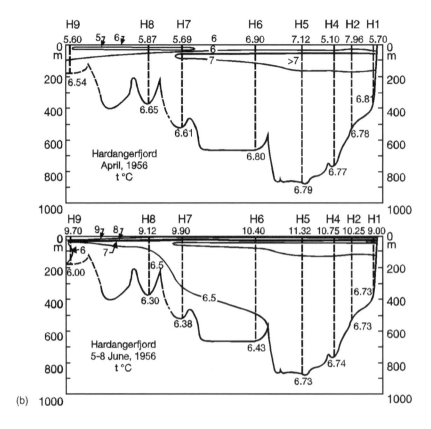

Figure 8.13 Distribution of (a) salinity and (b) temperature in Hardanger Fjord at various times. Reprinted with permission from Saelen (1967). Some features of the hydrography of Norwegian Fjords. In: *Estuaries* (Ed. G.H. Lauff). Copyright 1967 American Association for the Advancement of Science.

values, but little change has occurred in the temperatures further inland. It appears that there is a summer inflow of colder water, at sill depths, into the deep water of the fjord. This renewal may occur annually or less frequently and is probably independent of the upper layer circulation, being affected mainly by offshore water conditions.

Saelen (1967) used Knudsen's equations in analysing the results from Hardanger Fjord. The main difficulty of the approach is in estimating the depth of water involved in the ingoing bottom flow near the mouth, without measurements of the mean current velocity profile. It was estimated that the surface outflow at the mouth was between two to six times the river flow. Using Equation (6.21) the mean vertical velocity was calculated as $2–8 \times 10^{-6}$ m s^{-1}, corresponding to a daily vertical movement of about 0.5 m.

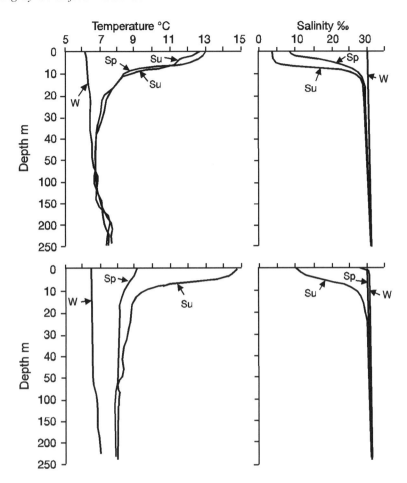

Figure 8.14 Representative seasonal temperature and salinity profiles for two stations in Knight Inlet. Upper panel, near the head, lower panel, near the mouth. Note the change of scale at 50 m depth. Reprinted from *Progress in Oceanography*, **12**, Farmer, D.M. and Freeland, H.J., The Physical Oceanography of Fjords, 147–219. Copyright 1983, with kind permission from Elsevier Science Ltd, The Boulevard, Langford Lane, Kidlington OX5 1GB, UK.

Knight Inlet, British Columbia

Farmer and Freeland (1983) have presented representative temperature and salinity profiles for two stations, one near the head and the other near the mouth. These profiles (Figure 8.14) show the changes that occur in stratification with different river discharge and meteorological conditions. High discharge occurs in the summer, with sharp stratification at the head with

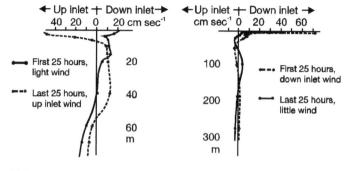

(a) Station 3½, July 6th to 8th, 1956   (b) Station 5, July 4th to 6th, 1956

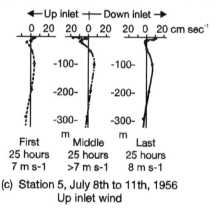

(c) Station 5, July 8th to 11th, 1956
Up inlet wind

Figure 8.15 Current measurements with varying wind conditions, Knight Inlet. Reproduced with permission of the Minister of Supply and Services Canada, 1996, from Pickard, G.L. and Rodgers, K. (1959) Current measurements in Knight Inlet, British Columbia. *J. Fish. Res. Bd. Can.* **16**, 635–678.

a surface layer of uniform salinity. In the winter stratification is almost nonexistent as the river flow is low.

Pickard and Rodgers (1959) have described extensive current measurements at two stations in the Inlet. Station $3\frac{1}{2}$ was situated on a sill with a depth of about 75 m and Station 5 was in a water depth of about 350 m in the inner basin of the Inlet. The maximum tidal range was about 5 m. Currents at all depths at station $3\frac{1}{2}$ showed tidal oscillations with peaks midway between predicted high and low water, i.e. a standing wave oscillation, but there were large irregularities in the currents at all depths. The amplitude of the tidal currents was about 0.50 m s^{-1} and changed little with depth. During the first 25 hours of measurement when the wind was light, there was a net outflow down to 40 m and a landward flow beneath. In the final 25-hour period with a 6 m s^{-1} up-inlet wind, the surface flow was landward to a depth of 6 m and seaward between 6 and 55 m (Figure 8.15a).

At station 5 the deeper current measurements showed tidal currents up to about 0.15 m s^{-1}. Slack water coincided with predicted high and low water, but there were again large irregularities in the flow. In the surface layer, however, the currents were several times larger, ranging between 1.20 m s^{-1} down-inlet to 0.45 m s^{-1} up-inlet. Strong surges in the current occurred in the hour before high water. At intermediate depths (10–50 m) the currents were irregular, sporadic and often zero. The flood current at 50–100 m started after predicted high water and extended down to 300 m at predicted low water. For the first 25 hours there was a down-inlet wind of 5 m s^{-1}, but during the last 25 hours there was little wind. The profiles of mean current (Figure 8.15b) show a faster, deeper surface flow with the down-inlet wind and a change in the mean flow at depth.

Further measurements at station 5 gave similar results, but in this case there was a continuous up-inlet wind which increased to 12 m s^{-1} at one stage. The mean current profiles (Figure 8.15c) show the increased wind speed reversing the surface current. When the gradient of the water surface was adjusted to the effect of the wind stress a normal current profile became re-established.

Deep Water Renewal

The deep water trapped inside the fjord may not be renewed very often, and in poorly ventilated fjords it can become oxygen-deficient very rapidly. To renew, the density of the water has to be higher at the mouth than within the basin, and this can happen on a range of time-scales from days to many years. There are four main processes involved.

1. *Tides*. Spring tidal mixing at the mouth can produce the inward transfer of deeper, more dense water on each flood tide (Russel Fjord; Reeburgh *et al.*, 1976), or when tidal currents are high (Puget Sound; Cannon and Laird, 1978), providing that overmixing is not produced. If this is the case, renewal may occur nearer to neap tides. Thus renewal can be on a fortnightly or a monthly time-scale.
2. *Meteorological forcing*. Down-estuary winds will enhance the entrainment between the brackish and intermediate layers, causing increased inflow at the sill that may involve denser water. Renewal is therefore intermittent.
3. *River discharge*. Increased discharge in the surface layer also creates enhanced mixing, though it also increases the thickness of that layer. In extreme cases, when the sill is shallow, complete blocking of the inflow is possible. Renewal is thus likely to be seasonal.
4. *Shelf processes*. The annual cycle of water density off the mouth will vary according to solar heating and the variation in the coastal current patterns. This can lead to renewal in late summer and autumn (Saanich Inlet: Anderson and Devol, 1973), in winter and spring (Skjomen: Skreslet and Loeng, 1977), though not necessarily every year.

Once denser water penetrates into the fjord it will act as a density current, flowing down the slope in the same way as described for the renewal in the Palmiet estuary. Some entrainment from the overlying water is likely so that the density of the current decreases as it travels. If it reaches a level at which it is neutrally buoyant it will separate from the bed and spread out as an internal layer of intermediate density. Thus the vertical structure of the deep water may be a complicated layering of water of different ages. The passage of a parcel of water in Puget Sound has been traced on its temperature and salinity characteristics by Ebbesmeyer *et al.* (1975).

Effects at the Mouth

Fjord circulation is composed of three layers, as shown in Figure 2.8. The surface layer of brackish water is maintained by river discharge and this varies little in thickness along the fjord, despite the entrainment of water from the intermediate layer below. This entrainment depends upon the tidal currents and wind stress. The tidal currents are largely confined to the surface layer and the interfacial shear and the entrainment varies tidally. The relative salinity difference between the layers decreases exponentially along the fjord and the salinity and surface layer thickness increase with increasing discharge. The salinity difference between the mouth and the head in the brackish layer is also often proportional to the salinity difference between the two layers at the mouth. Figure 8.16 shows the relation for Nordfjord, together with the monthly mean values, as the *fjord salinity loop*. The hysteresis may be due to transient response to fluctuating discharge.

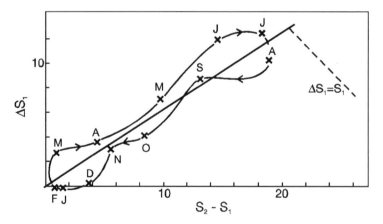

Figure 8.16 Fjord salinity loop for Nordfjord. The horizontal salinity difference between the mouth and the head ΔS_1 versus the difference between the salinity of the surface and bottom layers at the mouth $S_1 - S_2$. The crosses are monthly values. Reproduced from Stigebrandt (1981) by permission of Academic Press.

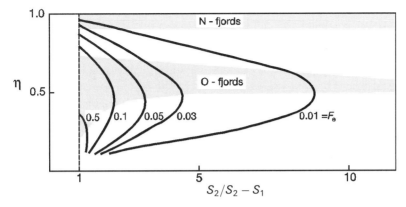

Figure 8.17 Fjord classification scheme in terms of estuarine Froude number. For N-fjords the interface is shallow with respect to the depth at the mouth. For O-fjords the two layers are of equal thickness. Reproduced from Stigebrandt (1981) by permission of Academic Press.

The thickness of the intermediate layer is limited by the sill depth at the mouth and the inflow is a result of the mixing of sea water into the brackish layer. Some of the mixing occurs at the sill itself and this is controlled, for small tidal velocities, by the internal Froude number of the layered flow at the mouth. In this situation the composite Froude number $G^2 = F_1^2 + F_2^2$ holds. When the layer thicknesses are equal, no matter how intense the mixing within the estuary, no increase in salinity of the outflowing mixture is possible. This process of *overmixing* appears to be an important control on the salinity of some harbours, such as New York and St John's, New Brunswick.

Stigebrandt (1981) has classified fjords according to the structure of the control section at the mouth as normal N, and overmixed O types. In N type fjords the interface is shallow with respect to the total depth, and only the surface layer is important in the hydraulics. For O type fjords the two layers are of approximately the same thickness, indicating a maximum mixing condition. Figure 8.17 shows the variation of relative depth of the two layers against the salinity difference $S_2/S_1 + S_2$. The curves of the estuarine Froude number F_e indicate how the salinity and thickness of the layers vary for a constant river discharge. The maximum value at a relative thickness of 0.5 shows the conditions when the fjord is overmixed and the vertical salinity difference cannot exceed the indicated value. When the river flow is small, and the value of F_e is small, the surface layer thickness is restricted, and the fjord is of the N type. This approach is only valid, however, for short restricted mouth conditions, where bed friction is small and where tidal effects are small.

The quasi-steady flow of stratified water over a shallow sill or through a constriction has been considered by Armi and Farmer (1986) and Farmer and Armi (1986) and reviewed by Largier (1992). This applies to both fjords and highly stratified, bar-built estuaries. There are three basic situations which depend on the composite Froude number G^2 and the estuarine Froude number F_e^2 at the sill (Figure 8.18).

1. *Blocked* — sea water does not intrude into the estuary (or fjord). This occurs if the net outflow in the fresh water creates a value of $F_1^2 = 1$ at the sill and $F_e^2 > 1$ over the sill.
2. *Exchanging* — the sea water intrudes as a bottom layer beneath an outflowing surface layer of fresher water. For this situation $G^2 = 1$.
3. *Plunging* — the sea water intrudes into the estuary, preventing the outflow of fresh water, but plunges beneath it somewhere upstream of the sill. Over the sill $F_2^2 = 1$ and $F_e^2 > 1$.

During the plunging a surface front is formed and this forms the *tidal intrusion front*. The actual position of the plunge point, or the blocking point, relative to the sill crest depends on the value of F_e^2.

Similar features occur for a constriction as for a sill, but the critical Froude number is then 0.3 rather than 1.0. Of course, both situations normally occur together.

When tidal effects are large the whole water mass moves across the sill and a variety of wave-like disturbances are created on the density interface, especially in the 'exchanging' situation. These have been described by Farmer and Freeland (1983) and depend on the thickness and characteristics of the layers. A schematic diagram of the response is shown in Figure 8.19. For a two-layer system over the sill lee waves are produced at maximum tide, there is blocking of the deep flow at the sill and the separation of a boundary layer beneath the trough of the lee wave downstream of the sill. When the tidal current decreases the lee waves progress upstream. For a three-layer flow waves are initially formed in the bottom interface, but these become suppressed as waves evolve on the near surface layering. These can produce mixing at maximum flow, but then the mixed water progresses upstream. For a long sill extensive mixing is produced, but this tends to collapse and spread out as the current reverses.

The internal wave drag created at the sill is the dominant way in which the tide loses energy in fjords. Sufficient energy is put into wave-like disturbances and hydraulic jumps to dominate the mixing processes and this can potentially happen at several sills within a fjord.

Tidal Intrusion Fronts

We have already seen that where the mouth of an estuary is constricted by a sill or a narrowing, that the brackish water can be blocked by the inflowing sea water which plunges beneath it. The plunge point is often visible as a change of colour, particularly if the brackish water is more turbid. The plunge line is also marked by a line of foam or debris which forms the surface convergence

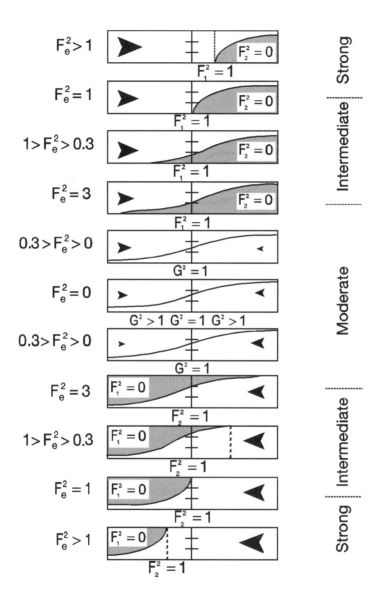

Figure 8.18 Density exchange flow through a constriction in terms of the estuarine Froude number F_e^2. Control is via the composite Froude number G^2 with negligible mixing and friction. The shaded layers are stationary. The arrows indicate the flow strength. Used by permission of the Estuarine Research Foundation. © Estuarine Research Foundation. Reproduced from Largier (1992).

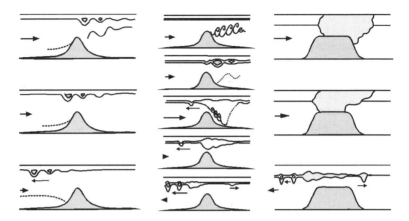

Figure 8.19 Flow structures over a sill at a fjord mouth. Left: schematic diagram of lee wave response to tidal flow over a sill. Upstream of the sill there is blocking of the deep flow. Downstream there is separation of the boundary layer beneath the trough of the lee wave. As the tide slackens the lee waves are released to form a wave train progressing upstream. Centre: schematic diagram of flow subcritical to mode 1, but critical to mode 2. In the early stages of the tidal flow the boundary layer separates from the sill crest, but subsequently separation is suppressed as mode 2 lee waves evolve. The lee wave collapses to spread out as a bore as the flow slackens. Right: sketch of tidal mixing over a longer sill leading to formation of water of an intermediate density that collapses and spreads out as an undular bore between the two layers as the flow slackens. Reprinted from *Progress in Oceanography*, **12**, Farmer, D.M. and Freeland, H.J., The Physical Oceanography of Fiords, 147–219, Copyright 1993, with kind permission from Elsevier Science Ltd, The Boulevard, Langford Lane, Kidlington OX5 1EB, UK.

between seaward flowing brackish water and landward flowing sea water. At the plunge line there is a downward flow in both layers, with a certain amount of entrainment of the fresher water into the more turbulent inflowing sea water.

These fronts were first described by Simpson and Nunes (1981), but there have since been many further cases described from other narrow estuaries. Figure 8.20 shows an example from the Esk estuary in north-east England. They are, of course, only formed on the flood tide.

Largier (1992) has defined three classes of tidal intrusion front (Figure 8.21) which relate strongly to the rate at which the estuary cross-section diverges upstream of the narrow section.

1. *Class I. Long-crested escarpment.* The plunge line follows the edge of an underwater escarpment, which is the limit of shallow water created by the deposition of sediment transported in through the mouth on the flood tide. The plunge line shape is governed by the bathymetry and the flow tends to approach the escarpment perpendicularly with little along-front shear.

2. *Class II. Weakly confined inflow.* The plunge line is primarily controlled by

Figure 8.20 A tidal intrusion front in the Esk estuary at the beginning of the flood tide. The salty clear water is flowing in from the left and under the turbid fresher water. Note the vortices shown by the foam at the tip of the intrusion. Photograph by the author.

the structure of the flow and the bathymetry is less important. Characteristically, the plunge line is U-shaped due to a parabolic cross-stream flow profile. Near the shore the slower flows plunge earlier and the faster midstream flows plunge later. The inertial terms and the pressure gradients are both important, but the flow is still largely normal to the front — there is no funnelling of the flow to mid-stream. Variations in this class depend on curvature in the channel and the influence of an underwater escarpment.

3. *Class III. Strongly confined flows.* The flow is confined to a narrow channel and there are strong cross-stream gradients due to friction on the sides. Inertia dominates immediately before the plunge point. The frontal shape is a V, which is little affected by the bathymetry. There is a significant flow component along the front which results in funnelling water and debris towards the point of the V where quite violent eddies are present in the brackish water, generally as contra-rotating pairs (Figure 8.20). Again, asymmetrical versions occur when the channel is not straight.

When classes II and III flow into a very wide area the plunge line can become separated from the walls, but there will be significant entrainment between the jet-like flow and the surrounding flow.

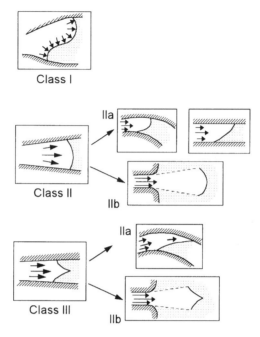

Figure 8.21 Idealized plan view sketches of classes and subclasses of tidal intrusion fronts. The sea is on the left and the estuary on the right. The shaded portion indicates fresh water. Used by permission of the Estuarine Research Foundation. © Estuarine Research Foundation. Reproduced from Largier (1992).

Plumes

When the Froude number at the mouth is less than critical, the brackish surface layer will be able to penetrate seawards of the mouth as a plume, especially during the ebb tide. Garvine (1977) states that a well-defined plume will exist beyond the mouth following the ebb tide when $P > 3/4$ and for all stages of the tide when $P > 2$ ($P = u_f/u_t$).

In highly stratified estuaries the buoyant surface layer flows out over the denser sea water, thins rapidly as it expands, being only a few metres thick at most, and can have a very high density contrast with the underlying sea water. An example is described for the Fraser River estuary by Yin *et al.* (1995). The contrast is also often visible as a colour or turbidity change. The dynamics is controlled by the pressure gradient created by the sloping fresh water surface, interfacial friction and entrainment across the frontal interface. The general structure is shown in Figure 8.22. The surface water flows down the slope towards the front. The interfacial friction slows the lower part of the layer and causes a return flow at the interface, where there is entrainment downwards into the lower layer. At the front there is a strong convergence as the depth

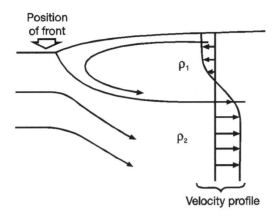

Figure 8.22 Schematic diagram through the edge of a plume.

mean flow in the plume carries it over the lower layer. The convergence rate appears to be larger in shallower plumes.

When the plume is laterally confined there is a thickened zone at the head of the plume where active mixing takes place by Kelvin–Helmholtz instabilities. This produces an intermediate layer in which the Richardson number is maintained at stable magnitudes. However, in a radially spreading plume there is likely to be high salinity gradients throughout, though entrainment occurs near the front.

The timing of the formation of the plume depends on the stage during the ebb tide when stratification or fresher water reaches the mouth. For short estuaries the plume appearance will be earlier than for longer ones. For well-mixed estuaries the intense bottom mixing creates a homogeneous layer that will thin rapidly and lift off the bottom as it exits the mouth. The salinity contrast will thus be controlled by that at the mouth and increases gradually during the ebb tide. If the offshore sea bed slope is small and the discharge large, inertia will delay the separation of the plume from the bed. Until then the fresher water will be bounded by inclined fronts reaching to the bed. Once it has left the mouth the plume movement will be controlled by the direction of the tidal currents offshore of the mouth, by wind and by the Coriolis force. On the up-tide side of the plume the front will be sharper and the plume thicker than on the downstream side and it is possible that the upstream side of the front may attain an equilibrium position where the tidal current and the spreading rate of the plume are equal. As the tide turns the plume can become separated from the estuary and advected alongshore, becoming thinner by further spreading. The major mixing will be carried out by surface wave action.

The structure and dynamics of plumes have been reviewed by Bowman (1988) and O'Donnell (1993).

Chapter 9

Partially Mixed and Well-mixed Estuaries

When tidal motion becomes significant in comparison with the river discharge, there is appreciable shear stress on the bed throughout the estuary. This changes the balance between internal mixing and mixing produced at the bed. There is interaction between the two mixing processes so that the tidally averaged salinity increases towards the bed and there is no homogeneous layer of near-bottom water. There is a continuous change in stratification with change in the tidal range, and because of the considerable variation in range with spring and neap tides it is difficult to separate partially mixed and well-mixed estuaries. An estuary may exhibit partially mixed properties and processes at some locations at the same time as showing well-mixed properties elsewhere. The Tamar estuary, for instance, shows much less stratification near its head where the water is shallow than further seaward where it is deeper.

In partially mixed estuaries the salinity stratification is sufficiently intense to limit the penetration of the bottom turbulence into the body of the flow, and there is a small amount of tidal frequency movement of the halocline. The dominant factor in the balance is the interaction between the surface barotropic tide and the steady horizontal density gradient (Jay and Smith, 1990b). As a result, partially mixed estuaries show the classical residual flow profiles produced by vertical gravitational circulation. There are still phase differences between the tidal oscillations at the bed and near the surface, but the form of the salinity profile remains similar throughout the tide as it is advected backwards and forwards. There is an asymmetry in the bed shear stresses and an internal tidal asymmetry similar to that in the surface tide. This is composed of two aspects: tidal straining and the fact that the barotropic and baroclinic gradients alternately oppose and reinforce each other on the ebb and flood tides. These are considered to be the primary mechanisms causing vertical shear in the residual flow.

As the estuary becomes better mixed the surface to bed phase differences become less marked and the gravitational circulation decreases in importance. The barotropic forcing then dominates. The vertical mixing is more intense and, though modified to some extent, the bottom induced turbulence reaches the surface. There is a smaller salinity difference between the surface and the bed, except near low slack water when stratification may appear, largely as a result of tidal straining. During the rest of the tide advection of the profiles dominates. It has been argued that these estuaries should be called weakly stratified as there is always some stratification at some time in the tide. Bowden (1981) has argued that the surface velocity of the density induced circulation needs to be only about 2% of the amplitude of the tidal current for the estuary to be no longer well mixed. Thus only small vertical salinity gradients are needed to produce partially mixed characteristics and the classical residual flow profiles.

In well-mixed estuaries the tide is normally strongly flood dominant and there is a large tidal trajectory of water movement and large velocity shears in both the vertical and the lateral directions. The longitudinal salinity gradient is fairly steep, with salt only penetrating for a few tidal excursions. However, the excursion can be long because of the high tidal range. There are also generally significant lateral gradients in salinity and these, coupled with the velocity shears, provide the main mechanism for the upstream dispersion of salt which is needed to counteract the downstream advection of the river flow. The velocity shear and the turbulence derived from the boundary friction extends throughout the water column and is little affected by small density gradients. The salt water is transported up the middle of the estuary on the flood tide and mixes with the fresher water on the sides and near the bottom because of the strong flood currents caused by the tidal asymmetry. On the ebb tide the currents are weaker, the shears are less intense and the saltier water near the boundaries are mixed in more slowly and thus carried seawards less effectively. On average this gives a dispersion of salt towards the head of the estuary. The mechanism is more effective when there are bays, harbours and other topographic features within which saltier water can become trapped on the flood tide, only to empty back into the ebb flood at a later stage in the ebb tide. This tidal trapping has been formalized by Okubo (1973). There are likely to be significant differences between narrow and wide estuaries, as the latter does not have the same intensity of lateral shears. However, the wider estuaries can develop horizontal circulations which are manifested by considerable differences across the estuary in mean flow and in salinity. This distinction has been investigated by Jay (1990). In a narrow estuary the Lagrangian mean transport on a cross-section is approximately equal to the river flow. This is because the inward Stokes drift is exactly compensated by an outward Eulerian flow, which, together with the river flow, is the non-tidal drift. In wide estuaries this is not necessarily so as the compensation flow can be on one side of the channel rather than equally distributed over the cross-section, thereby causing horizontal circulation. Thus it is possible to have an upstream net salt flux on one side of the estuary and a net downstream flux on the other, though, obviously, the cross-sectional average has to balance in the same way as it does in narrow estuaries.

Studies of the longitudinal dynamic balance in well-mixed estuaries has shown that there is mainly a balance between the surface water slope, the barotropic force and the bed shear stress. The slight imbalance then creates the time and spatial accelerations. The studies have shown that there is an asymmetry in the drag coefficient during the tide (Brown and Trask, 1980). This may be the result of a small vertical density structure, so that the flow is not completely homogeneous. Energy is then dissipated within the flow as well as because of bottom stress. Consequently, a single value of the drag coefficient is inappropriate for accurately modelling the flow in well-mixed estuaries.

There are no precise boundaries between the two classes and the same processes operate in both, it is just a matter of degree. Consequently, they are described here together.

JAMES RIVER, VIRGINIA

Extensive measurements were undertaken in 1950 on the James River estuary and these formed the basis of the classical analysis of Pritchard (1952a, 1954, 1956, 1967), which emphasizes the tidally averaged structure. The section of the estuary investigated was between 20 and 45 km above the mouth (Figure 9.1). A number of stations were occupied for three periods of at least four days, during which serial measurements of current, salinity and temperature were taken. Several stations were concentrated on three cross-sectional traverses. The velocities were obtained using a current drag (Pritchard and Burt, 1951). The surface high and low water salinities for one day are shown in Figure 9.2. The surface salinities are noticeably lower on the right-hand side. At other times similar distributions were observed, with surface salinities at high water being about 4 higher than those at low water. The mean vertical salinity profiles at station J17 show a mid-depth halocline and a difference of about 2 between the surface and lower layers (Figure 9.3). The mean velocities show a two-layer flow with an upstream flow in the bottom layer and a downstream flow in the surface layer. The mean of the seaward velocity in the surface layer was about 19% of the average tidal velocities, whereas in the lower layer the landward flow was about 20% of the average tidal velocities. The mean flow summed over depth was much larger than that due to the seaward movement of river flow. At station J17 the level of no net motion sloped by over 1 m downwards towards the right-hand side of the estuary.

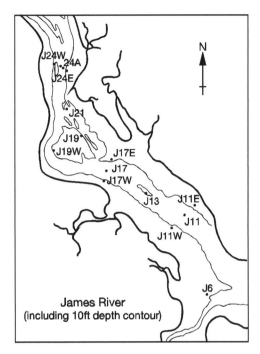

Figure 9.1 James River estuary, topography and station postions. From Pritchard and Kent (1953). Reproduced with permission.

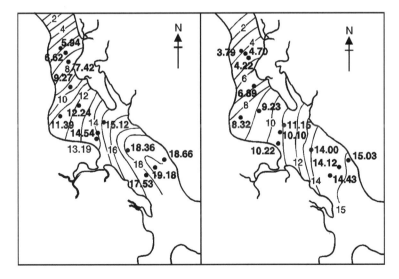

Figure 9.2 Surface salinity distributions in the James River estuary. Left-hand panel, high water; right-hand panel, low water. From Pritchard (1952b). Reproduced with permission.

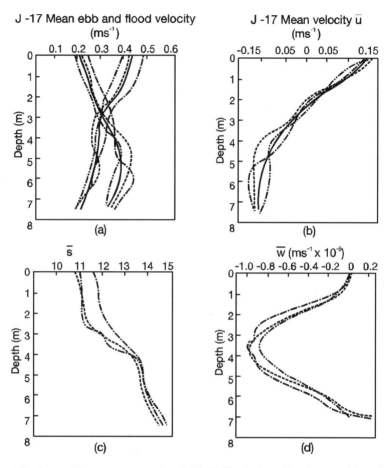

Figure 9.3 James River estuary, station J-17. (a) Vertical profile of mean ebb and flood currents for three different periods. Full line, weighted average of all three periods. Net flow is seawards in the upper layer and up-estuary in the lower layer. (b) Vertical profile of tidal average velocity for three periods. Full line, weighted average of all three periods. (c) Vertical profile of mean salinity. (d) Mean vertical velocity. From Pritchard and Kent (1953). Reproduced with permission.

Using the differences in mean horizontal velocity between the three cross-sections, mean vertical velocities were calculated. The vertical velocities calculated at station J17 are upward at all depths with a maximum value of about $1 \times 10^{-5}\,\mathrm{m\,s^{-1}}$ at mid-depth (Figure 9.3).

This set of observations is probably the most extensive and the most thoroughly analysed from any estuary. Many other investigations have been greatly influenced by this study. The reasons for these typical features should be clear from the explanations elsewhere in this book.

Pritchard and Kent (1953) and Pritchard (1954) have considered the salt balance in the James River. The estuary was assumed to be laterally homogeneous and comprised a series of elements of equal thickness, but whose width b varied with depth and along the estuary between the cross-sections 11, 17, 24.

Equations (5.2) and (6.9) for the continuity of water and of salt were used in this case. The tidal fluctuation terms were considered negligible. The time change of mean salinity was also small. The vertical velocities were calculated using Equation (6.7) and the two advective terms were then calculated from the observed data. The cross-sectional mean value of the horizontal diffusive flux $(\overline{u's'})$ was calculated as less than 5% of either of the mean advection. The vertical non-advective term was found by simple addition, which, of course, included computational errors.

The horizontal advective term and the vertical non-advective (turbulent) term were the most important. The vertical advective term became important at mid-depth near the halocline. In spite of the possible errors involved in the computation of the values for horizontal diffusion the balance seems intuitively correct. The vertical distribution of salinity represents a balance between vertical eddy diffusion, the horizontal advection and the vertical advection. No account has been taken of possible lateral effects and secondary circulations.

Assuming zero values of eddy diffusion at the surface, integration of the vertical diffusion data shows that the vertical non-advective flux of salt was a maximum at mid-depth (Figure 9.4a). The vertical eddy diffusion coefficient K_z calculated from these results reached $9\,\mathrm{cm}^2\,\mathrm{s}^{-1}$ (Figure 9.4b). As tidal movements are the most important factor governing the mixing processes there should be a relationship between the tidal current velocities and the vertical non-advective flux of salt. Pritchard demonstrates this for three periods of observation at three sections in the James River, the salt flux rising with the tidal velocities.

MERSEY NARROWS, ENGLAND

The Mersey Narrows is a section of the River Mersey about 10 km long and 1 km wide connecting Liverpool Bay with a wider and shallower inner estuary (Figure 9.5). The maximum depth in the Narrows exceeds 20 m. The tide range can reach 9.4 m and produces currents up to $2.2\,\mathrm{m\,s}^{-1}$. The river flow is small compared with the large tidal flow, but is very variable, ranging from about $25\,\mathrm{m}^3\,\mathrm{s}^{-1}$ to over $200\,\mathrm{m}^3\,\mathrm{s}^{-1}$. The estuary has been described by Hughes (1958) and Bowden and Sharaf el Din (1966a).

The longitudinal variations of salinity at high and low water are shown in Figure 9.6. The difference between the extreme values of salinity at any position, during a tidal cycle, was about 4 for a low river discharge period. For high river discharge the difference was about 11. During low river

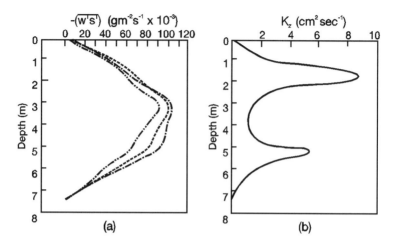

Figure 9.4 James River estuary, station J-17. (a) Vertical turbulent flux of salt as a function of depth. (b) Vertical eddy diffusivity K_z. From Pritchard and Kent (1953) and Pritchard (1967). Reproduced with permission from the American Academy for the Advancement of Science.

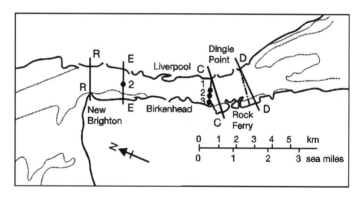

Figure 9.5 Narrows of the Mersey estuary showing positions of stations and cross-sections. Dotted lines, low water spring tides. Reproduced by permission of Blackwell Science Ltd from Bowden and Sharaf el Din (1966a).

discharge the salinity differences between the surface and the bottom were not a maximum at high or low water, but continued to increase to about 1 about 40 minutes after high water and to more than 1 over an hour after low water. The variation of salinity with depth was small or absent when the tidal current was strong, so that for the mid-tide period the isohalines were almost vertical. For high river discharge a vertical salinity gradient was always apparent and there was no period when the isohalines were vertical, also the differences in surface to bottom salinities at slack water were larger than for low river discharge.

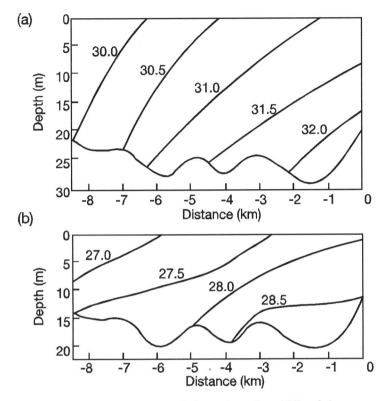

Figure 9.6 Narrows of the Mersey. Isohalines along the middle of the estuary at (a) high water and (b) low water. Distances from New Brighton. Reproduced by permission of Blackwell Science Ltd from Hughes (1958).

Low salinity surface waters are likely to exist well into Liverpool Bay and offshore observations of currents and salinity have shown that an estuarine type circulation extends to a distance of at least 18 km from the mouth of the Mersey (Bowden and Sharaf el Din, 1966b).

Current and salinity measurements at stations on section C have been reported by Bowden and Sharaf el Din (1966a) and at stations along the Narrows by Bowden (1963). The latter measurements showed a surface outflow and a bottom inflow, but with variations in the depth-mean of the residual flow between 0.01 to $-0.087\,\mathrm{m\,s^{-1}}$. These values are too large to represent the mean flow through a complete section and were probably caused by lateral variations. At each station there was a downstream flow near the surface and an upstream flow near the bottom, relative to the depth mean flow. The former measurements showed a depth mean flow of $0.077\,\mathrm{m\,s^{-1}}$ upstream, $0.09\,\mathrm{m\,s^{-1}}$ downstream and $0.104\,\mathrm{m\,s^{-1}}$ downstream at stations 1–3, respectively (Figure 9.7). The lateral mean velocities indicated a considerable

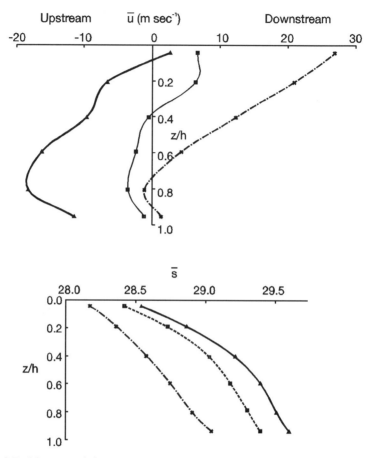

Figure 9.7 Narrows of the Mersey. Mean velocity and salinity at stations on section C. Triangles, station 1; squares, station 2; crosses, station 3. Reproduced by permission of Blackwell Science Ltd from Bowden and Sharaf el Din (1966a).

eastward flow at all stations which may be associated with the curvature of the channel near this section. There were indications, however, of an anticlockwise secondary circulation westwards on the surface and eastwards near the bottom. The mean salinity profiles for this period (Figure 9.7) showed that the fresher water was concentrated on the left-hand side of the section.

Salinity observations over the period October 1962 to October 1963 were completed at a number of stations within the Narrows (Bowden and Sharaf el Din, 1966a). It was found that the best correlation between the salinities and the river flow occurred using the average river flow for the week ending on the day of salinity sampling. This is obviously a measure of the residence time of the fresh water within the estuary.

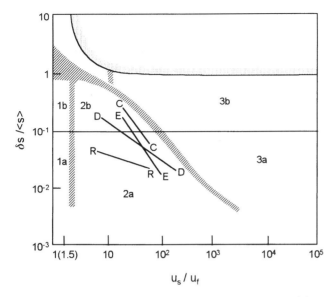

Figure 9.8 Stratification–circulation diagram for the Mersey. Positions of the cross-sections are shown on Figure 9.5. From Bowden and Gilligan (1971). Reproduced with permission from the American Society of Limnology and Oceanography.

The horizontal and vertical gradients of salinity depend on the river flow more than on tidal range. However, Bowden and Gilligan (1971) found that the density current transport was relatively insensitive to changes in river discharge, but increased with increasing tidal currents. Similar results have been found for suspended sediment transport (Price and Kendrick, 1963), the transport increasing rapidly with tidal amplitude.

If we assume, following Pritchard's work in the James River, that horizontal advection and vertical eddy diffusion are the dominant processes affecting the distribution of salinity, and that this is applicable to observed as well as mean values, then estimates of the vertical eddy diffusion coefficient can be made (Bowden 1960, 1963), as outlined in Chapter 6.

From extensive measurements on four sections in the Mersey Narrows, Bowden and Gilligan (1971) have calculated $\delta s/\langle s \rangle$ and u_s/u_f and plotted the results on a stratification–circulation diagram (Figure 9.8). The results for each section show considerable scatter, possibly due to non-steady state conditions operating, and also lateral variations in excess of those allowed for in the calculations. However, lines through the points lie almost parallel to each other, yet are displaced from each other. All of the lines lie within the type 2 classification in which the net flow reverses with depth and both diffusion and advection are important in the upstream flux of salt. The furthest seaward section lies entirely within the 2a category where the stratification is relatively

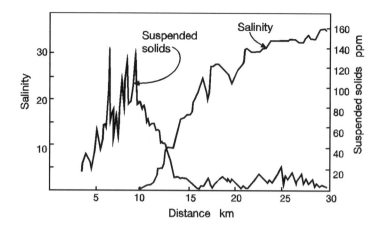

Figure 9.9 Distribution of salinity and suspended solids on a traverse at about high water up the Tamar estuary showing the turbidity maximum close to the head of the salt intrusion. Distances are from the tidal limit.

slight. The other sections extend into the 2b region where the stratification is greater.

TAMAR ESTUARY, ENGLAND

The Tamar estuary is a mesotidal/macrotidal, partially mixed, flood dominant estuary. About 30 km long, it has an average river discharge of $22\,m^3\,s^{-1}$ and a tidal range of 2.1 m at neap tides and 4.7 m at springs. By topography it would be classified as a ria. Its characteristics have been described in a series of papers (Uncles and Stephens 1989; Uncles *et al.*, 1983, 1984, 1985a, 1985b, 1988). The axial distribution of salinity is shown in Figure 9.9.

The situation within a tide is shown in Figure 9.10 for a station near the head of the estuary. At neap tides the flood currents reached a maximum of about $0.6\,m\,s^{-1}$ at a time when advection brought the head of the salt intrusion up the estuary to that location. The high current speeds ensured that the water column was well mixed at that time. Also, the suspended sediment concentration reached $500\,mg\,l^{-1}$ as a result of combined advection and entrainment of the sediment off the bed. The concentrations dropped rapidly within the salt intrusion and as the flood velocities diminished. Consequently, a turbidity maximum was present. The movement of the concentration contours towards the bed indicated that there was deposition of sediment at the high water position of the turbidity maximum. At high water the surface water started to ebb almost three hours before that at the bed. Consequently, surface currents reached $0.4\,m\,s^{-1}$ when the water at the bed was stationary. This shear (which is an expression of tidal straining) carried water of decreasing salinity out over water of almost constant salinity and created high stratification until the salt intrusion disappeared. During this internal shearing period entrainment eroded the salt intrusion at a rate of about $4 \times 10^{-4}\,\mathrm{m\,s^{-1}}$ and its disappearance was mainly the result of interfacial erosion with little advection. The turbidity maximum then reappeared as a consequence of advection from upstream, with some reentrainment of sediment from the bed. Averages over the tidal cycle gave the classical salinity and velocity profiles.

Similar events occurred during the spring tide. The current velocities on the flood tide, however, increased to almost $1\,\mathrm{m\,s^{-1}}$, with a consequential increase in suspended sediment concentration to almost $1\,\mathrm{g\,l^{-1}}$. The flood currents lasted a shorter time than on neaps. The increased tidal trajectory meant that advection up the estuary increased the maximum bed salinity at high water, but there was still the period of internal shearing at the beginning of the ebb tide. The ebb currents were similar in magnitude to those on the neap tide, but lasted longer.

As far as the suspended sediment is concerned the flux (concentration times the velocity) over the flood tide was much greater than that over the ebb tide. Thus it is obvious that a large amount of sediment must have been deposited over the high water landward of this station. During periods of increased river flow the balance during the tide changes, with the ebb currents becoming relatively stronger, and sediment is transported down the estuary, only to be transported landwards again when flow decreases. Thus there is a seasonal transfer of fine sediment up and down the estuary.

CONWY ESTUARY, NORTH WALES

The estuary is located in North Wales (Figure 9.11) and is about 23 km long, with a spring tidal range of 6.9 m. Freshwater discharge has a mean value of about $20\,\mathrm{m^3\,s^{-1}}$, but varies over the range 1–$500\,\mathrm{m^3\,s^{-1}}$. As the estuary is narrow, being only a few hundred metres wide over much of its length, the tidal currents reach over $1\,\mathrm{m\,s^{-1}}$. The tidal conditions have been described by Knight (1981). The estuary is flood dominant, the flood current being about 50% greater than the ebb tide almost irrespective of the tidal range. The flood currents are reduced and the ebb currents increased by increases in the river discharge, however. Figure 9.12 shows the change in time and range of the tide along the estuary. A typical tidal variation of salinity along the estuary for low river flow (Figure 9.13) shows a tidal salinity excursion of about 10 km. The vertical and transverse salinity differences have been examined for the Tal-y-Cafn reach by Guymer and West (1991). They show that the vertical salinity differences decrease to less than unity during the flood tide and remain low over high water and during the ebb tide. The transverse salinity differences are of a similar magnitude and show similar patterns of change. Transverse salinity differences on the flood tide are greater for spring tides, with a trend for more

148

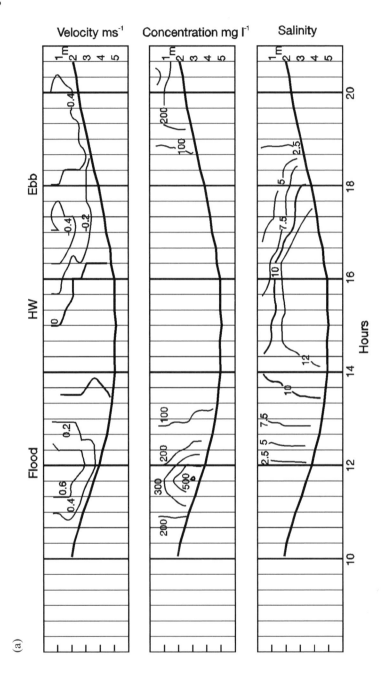

(a)

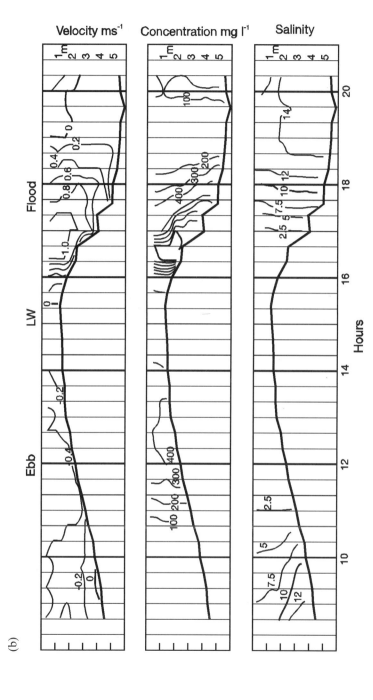

Figure 9.10 Isolines of velocity, suspended solids concentration and salinity for a station near the head of the Tamar estuary. (a) Neap tide and (b) spring tide. The water surface is taken as the origin for the water depth.

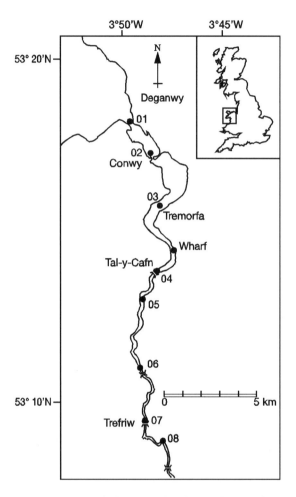

Figure 9.11   Conwy estuary, North Wales showing positions of the stations.

dense water towards the centre and the bed of the channel. This is consistent with the observed axial convergence line and a twin-cell secondary circulation (Nunes and Simpson, 1985).

Knight (1981) has used the measured surface water slopes, currents and density distribution in the longitudinal equation of motion to determine the bed shear stresses. He found that the resistance coefficients and the bed roughness length varied considerably during the tide and that a single value could not be used in a one-dimensional model to represent tidal movement, in agreement with the studies mentioned earlier.

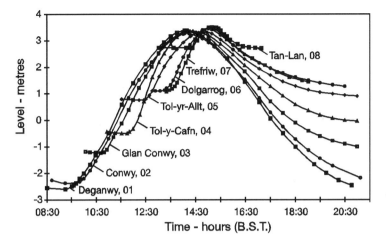

Figure 9.12 Water level curves for a 8.8 m tide in the Conwy estuary. For station positions, see Figure 9.11. Reproduced by permission of Academic Press from Knight (1981).

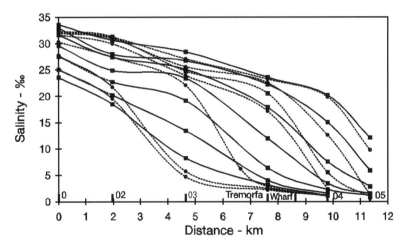

Figure 9.13 Instantaneous longitudinal salinity profiles for a 7.5 m tide in the Conwy estuary. Crosses, ebb tide and circles, flood tide. For station positions, see Figure 9.11. Reproduced by permission of Academic Press from Knight (1981).

The movement of saline water within the estuary has been examined by Turrell *et al.* (1996) during a period when river flow was about $1.5\,\text{m}^3\,\text{s}^{-1}$. There was a large ebb–flood asymmetry in the temporal salinity changes. Brackish water filled all but the lower estuary at low water, but as the flood tide commenced a zone of high salinity gradient was advected inland. The longitudinal gradient diminished around high water, but gradually increased again during the ebb tide, though it remained only a quarter to a half of the magnitude of that on the flood. Thus the flood tide recharges the axial salinity gradient with new sea water and the ebb tide largely removes the mixed water, replacing it with fresh river water. There was an overall increase in the low water salinity as the tide range increased towards spring tides.

The Tal-y-Cafn reach has been used for extensive dye diffusion studies by West and Cotton (1981) and Guymer and West (1988). The former workers found that the transverse diffusion coefficient, though it included secondary flow effects, agreed with laboratory data and may be given by $K_y = 0.4(hu_*)^{1.12}$. The coefficient for the vertical diffusion coefficient (i.e. K_z/hu_*) varied between 0.01 and 0.025 for ebb and flood tides, respectively.

Guymer and West (1988) found that K_y averaged $270\,\text{cm}^2\,\text{s}^{-1}$ for the ebb tide and $840\,\text{cm}^2\,\text{s}^{-1}$ for the flood tide, though the latter included the effect of secondary circulation. Values for K_Z on the ebb tide were $6\text{--}21\,\text{cm}^2\,\text{s}^{-1}$.

VERTICAL GRAVITATIONAL CIRCULATION

As we have seen in Chapter 7, the longitudinal pressures that drive vertical gravitational circulation are composed of contributions from the surface water slope (the barotropic component) and the density field (the baroclinic component). The former gives pressure gradients that are constant with depth, whereas the latter creates horizontal pressure gradients that are zero at the surface and increase to a maximum at the bed. On the ebb tide the two pressure fields oppose each other (Figure 9.14). Near the surface the barotropic exceeds the baroclinic, relative to the depth mean, and near the bed the reverse is true. Thus there is a relative outward force near the surface and an inward force near the bed. On the flood tide both forces act inwards and the magnitude is still inwards near the bed and outwards near the surface, relative to the depth mean. Taking the mean over the tidal cycle, in the upper layer the surfaces of equal pressure slope downwards towards the sea and maintain the residual flow towards the sea. Near the bed the pressure surfaces slope downwards towards the head and maintain a landwards residual flow. Near mid-depth the surfaces are horizontal, the pressure gradients are zero and this coincides with the depth of no motion.

The strength of the vertical gravitational circulation is a balance between the horizontal density gradient and the vertical mixing, as defined by the vertical eddy viscosity. This balance is shown by inspection of Equation (9.1). During the tide vertical mixing is produced by the turbulence generated by the tidal motion and this affects the vertical stratification. The horizontal density field is affected to a lesser extent, remaining relatively constant during the tide. Nevertheless, the changes that occur in the horizontal density gradients are sufficient to alter the intensity of the pressure field and the relative residual

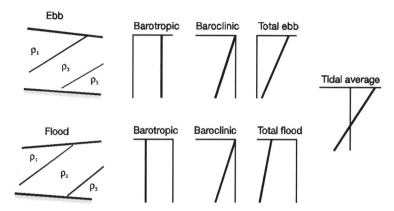

Figure 9.14 Diagrammatic balances between barotropic and baroclinic forces on an ebb tide (top) and a flood tide (bottom) and the tidal average.

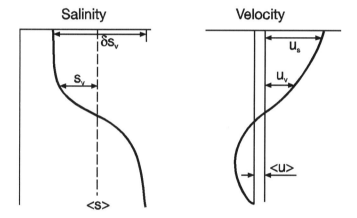

Figure 9.15 Definition sketch for the analysis of salt fluxes.

velocities of the two layers. An increase in the tidal velocities will cause vertical mixing, which decreases the vertical eddy viscosity, but which also decreases the horizontal density gradients. The overall effect is to reduce the vertical gravitational circulation. When the tidal velocities are less, the horizontal gradients are increased and the vertical stratification is increased, thus reducing the vertical exchange and increasing the eddy viscosity. Again the effect would be to reduce the vertical gravitational circulation. Thus there is an optimum strength for the gravitational circulation in the partially mixed situation.

Officer (1976) has characterized the residual profiles of velocity and salinity from a theoretical standpoint, using the terminology in Figure 9.15, as

$$u_v = \frac{1}{48} g \frac{\partial \rho}{\partial x} \cdot \frac{h^3}{\rho N_z} (8\eta^3 - 9\eta^2 + 1) \qquad (9.1)$$

where $\eta = z/h$.

Thus for the mean surface velocity u_s where $\eta = 0$

$$u_s = \frac{1}{48} g \frac{\partial \rho}{\partial x} \frac{h^3}{\rho N_z}$$

Also the salinity deviation s_v from the depth average $\langle s \rangle$ is

$$s_v = u_s \frac{h^2}{K_z} \frac{\partial s}{\partial x} \left(\frac{2}{5} \eta^5 - \frac{3}{4} \eta^4 + \frac{1}{2} \eta^2 - \frac{1}{12} \right) \qquad (9.2)$$

This gives the difference in salinity δs_v between the bed and the surface as

$$\delta s_v = \frac{3}{20} u_s \frac{h^2}{K_z} \frac{\partial s}{\partial x} \qquad (9.3)$$

The salt flux Q through the water column will be the integral over depth and over the tide of the product of the velocity and the salinity. This gives

$$Q = h\langle \bar{u} \rangle \langle \bar{s} \rangle + h \overline{US} + h\langle \overline{u_v s_v} \rangle + \text{turbulent fluxes}$$
$$ S_1 \qquad\quad S_2 \qquad\quad S_3$$

The first term S_1 is the flux due to advection on the river discharge. S_2 is called tidal dispersion and S_3 is the flux on the gravitational circulation.

For a steady state,

$$S_1 + S_2 + S_3 = 0 \qquad (9.4)$$

From Equations (9.1) and (9.2), and integrating through the depth

$$S_3 = -0.030 \frac{u_s^2 h^2}{K_z} \cdot \frac{\partial s}{\partial x} \qquad (9.5)$$

The fraction of the salt flux due to dispersion $v = S_2/S_1 = 1 + S_3/S_1$

From Equations (9.4), (9.5) and (9.3)

$$\frac{u_s h^2}{K_z} \frac{\partial s}{\partial x} = \frac{20}{3} \delta s_v$$

where δs_v is the difference between surface and bed salinities.

Therefore $S_3 = -0.030 \, u_s \cdot 20/3 \, \delta s_v$ and

$$v = 1 - 0.20 \frac{u_s}{\langle u \rangle} \frac{\delta s_v}{\langle s \rangle} \qquad (9.6)$$

Thus when tidal dispersion is a large fraction of the total salt flux, the gravitational circulation is small. Conversely, when gravitational circulation is large, the contribution to the flux due to the tidal dispersion is small. Note, however, that the tidal variation in water depth is neglected. This is an important constraint for shallow estuaries. This analysis gives the tidal dispersion fraction in terms of two dimensionless parameters that are the circulation and stratification parameters of Hansen and Rattray (1966). Contours of the value of v plotted on their classification diagram (Figure 2.14b) show that partially mixed estuaries have both tidal dispersion and gravitational circulation, with values of v between 0.1 and 0.9. Further consideration of the values of tidal dispersion fraction is given elsewhere in this book.

TIDAL STRAINING

As we have seen for the station on the Tamar, there is considerable variation of stratification during the tide. This is caused by *tidal straining*. Because of shear in the velocity profile, the surface water travels faster and with a larger trajectory of motion than the near-bed water. On the ebb tide the fresher water is thus differentially displaced over the saltier water below and the stratification increases. Thus there is a faster decrease of surface water salinity than near the bed. The surface to bed salinity difference increases, but may briefly decrease at maximum current by mixing, only to increase again towards low water. On the flood tide the differential movement will tend to diminish the stratification and even without mixing the stratification can be eliminated by the time of high water. This has been called *overstraining* (Nepf and Geyer, 1996) and would notionally lead to the situation of denser water being advected over less dense water if it were not for the mixing that inevitably results.

Tidal straining can thus be a main control on the strength of the stratification produced by the input of buoyancy by the fresh water. The degree of stratification produced will depend on the horizontal density gradient and on the tidal excursion. Simpson *et al.* (1990) have determined a criterion for the strain-induced periodic stratification by considering the average input of buoyancy over the ebb tide by straining, in comparison with the mean tidal stirring power over the same period. Periodic stratification will occur when

$$\frac{1}{\rho}\frac{\partial \rho}{\partial x} > 2.2 \times 10^5 \left(\frac{u_{\text{ebb}}}{h}\right)^2$$

where u_{ebb} is the average velocity over the ebb tide and h the water depth.

Sharples *et al.* (1994) have applied the principle to the Upper York River estuary and found that the stratification changes could only be successfully modelled if the horizontal density gradient was not constant with depth. The estuary was ebb dominant and the stratification changes were sensitive to the degree of mixing occurring on the ebb currents, with the result that towards spring tides the estuary tended to become well mixed rather abruptly when the tidal range exceeded a critical value. These *destratification events* had previously been observed by Hass (1977). The York River estuary, being ebb dominant, had very little phase difference through the water column in the turn of the current at high water, but a difference of almost an hour at low water. This obviously assisted the flood tide in creating the stratification destroyed on the ebb tide. This makes an interesting contrast with the flood dominant Tamar estuary, where there is significant shearing at high water and vertical mixing is greatest on the flood tide.

In the Hudson River Nepf and Geyer (1996) found that the tidal straining accounted for the extent and timing of mixing during the tide, estimated from the distribution of Richardson number. Comparing the flood and ebb tides the regions of mixing were structurally different even though their extents were similar. During the flood active mixing occurred in a well-mixed near-bed layer, whereas on the ebb mixing occurred throughout regions of significant stratification. This difference was due to tidal straining on the ebb tending to maintain stratification despite mixing, while on the flood the straining tended to reduce stratification and thereby enhance the effects of mixing near the bed. Consequently, the maximum and minimum near-bed stratification occurred during late ebb and late flood, respectively.

EFFECT OF WEATHER ON ESTUARIES

If we carry out short-term measurements within an estuary, say a tidal cycle or two, the classical mean velocity profiles may not be observed. The reason for this is generally the effects of the weather, particularly the wind. The first major study of this was a year-long series of current measurements in the Potomac Estuary by Elliott (1978). After removing the tidal velocity fluctuations he found longer period fluctuations in the residual currents of a period between two and five days and with an amplitude of about $20\,\mathrm{cm\,s^{-1}}$. This agreed with the passage time for atmospheric depressions. Six different circulation patterns were defined. The classical circulation of a downstream surface flow and an upstream bottom flow, though the most frequently observed pattern, only occurred for 43% of the time, with a mean duration of 2.5 days. A downstream wind enhanced the surface outflow and increased the vertical mixing and the bottom inflow. The reverse, with inflow on the surface and outflow on the bed, occurred for 21% of the time. This occurred when the wind was blowing into the estuary. Total inflow, or storage, which implies an increase in the estuary volume, occurred for 22% of the time. The balance of the time was accounted for by total outflow or by three-layer circulations. There was a sequence in the events that was accounted for by the progression of winds around passing depressions. The total outflow and storage periods occurred as a result of interaction of the Potomac with Chesapeake Bay and the local coastal waters (Elliott and Wang, 1978). A wind blowing towards the south-west down the coast causes water levels to rise towards the coast, as a result of *Ekman transport*, which would raise the water level in the estuaries. Conversely, a wind from the south-west would cause offshore Ekman transport and a lowering of the water level in the estuaries. Similar wind effects have been observed in many other estuaries and some modelling of the effects have been carried out by Garvine (1985). Van de Kreeke and Robaczewska (1989) show for the Volkerak estuary that a down-estuary wind increases the vertical exchange of momentum and decreases the gravitational circulation. Obviously the observed changes are unlikely to be uniform across the estuary and some compensatory adjustment is likely in the shallow water on the sides of the estuary.

Wind is also an important influence on the surface layer of fjords. Pickard and Rodgers (1959) showed that up-inlet winds could reverse the surface layer flow, presumably until the surface slope was established to oppose the wind stress. Outflow during this time was concentrated at the base of the surface layer. A down-inlet wind created an enhanced surface outflow and a compensating inflow between 40 and 100 m. Similar observations are reported by Farmer and Osborn (1976).

The continual effect of the wind and river discharge on estuarine circulation leaves us with a major problem in trying to determine the fluxes within estuaries, and particularly the fluxes from estuaries into the coastal areas. To measure the flux of a pollutant from an estuary requires a statistical approach that includes a reasonable range of the weather conditions. Weisberg (1976) has examined a time series from Narragansett Bay and shows that an averaging time in excess of ten days is required to obtain an estimate of the mean flow unaffected by wind. Consequently, measurements taken over only a few tidal cycles are unlikely to be representative and there is the possibility that an estuary is seldom in steady state. The time change in the mean salinity cannot therefore be considered negligible. This means that short-term measurements are of little help in solving some significant estuarine problems and we need to resort to mathematical models.

TURBIDITY MAXIMUM

One of the most distinctive features in partially and well-mixed estuaries is the presence of a turbidity maximum. This is a zone of high concentrations of suspended sediment, higher than in the river or lower down the estuary, and which is located at, or near to, the head of the salt intrusion. For a general review, see Dyer (1995b).

The peak concentration of suspended sediment varies considerably with the tidal conditions. Despite differences due to sediment availability, mesotidal estuaries have maxima with concentrations of the order of 100–200 ppm $(mg\,l^{-1})$, whereas hypertidal estuaries have concentrations of the order of 1000–10 000 ppm $(mg\,l^{-1})$. The turbidity maximum contains a high proportion of a narrow size range of mobile fine sediment, and the maximum can contain a mass of sediment several times the amount brought down annually by the river. Consequently, discharge of the particles to the sea must be restricted, the particles are trapped and their residence time is likely to be well in excess of a year. The sediment involved is cohesive and the particles flocculate, forming larger aggregates. As their size increases, their density decreases, but their settling velocities increase. Their size is governed by the suspended particle concentration, the turbulent shearing and such factors as salinity and organic content. Thus in the turbidity maximum there is an interaction between the fluid dynamics and the suspended sediment properties.

The position of the maximum varies with river discharge, with the maximum moving downstream with increasing flow (Uncles and Stephens, 1989). The total mass in the maximum also increases. However, the movement downstream involves expansion into an increased cross-sectional volume. Thus the concentrations could decrease even though the mass increases.

The turbidity maximum also moves up and down the estuary during the tide and the concentrations alter as it does so because of the entrainment of sediment from the bottom and settling. At about high water the maximum is located well up the estuary and concentrations are relatively low because of settling. During the ebb tide the maximum is advected down the estuary and becomes longer, with higher concentrations, because of the erosion of sediment from the bed. Settling again occurs at around low water, with further erosion on the flood tide. Thus there is a cycling of sediment occurring between the bed and the water on a regular basis.

The mass of sediment in suspension also varies between spring and neap tides. The higher bed shear stresses during spring tides are capable of eroding more sediment and the more energetic turbulence is able to keep it in suspension. For instance, in the Severn estuary the mean total mass in suspension varies by $5\text{--}30\,\mathrm{kg\,m^{-3}}$ between neap and spring tides. On the decreasing tides the sediment is progressively deposited and not all of it is re-eroded on the following tide, leading to accumulation of rather soft mud in the channel, in the intertidal areas and in low velocity areas, such as docks.

On a tidally averaged basis the location of the turbidity maximum at spring tides is further up the estuary than at neap tides. This is due to the increased mean water level at the head of the estuary caused by the increased range at spring tides. An increased volume of water is present at high water, but only a small decrease occurs at low tide, leading to a relative increase in the mean water level. Spring tides enhance the deposition of sediment in the intertidal areas at the head of the estuary due to the long slack water period over high water, together with the high concentrations.

Associated with the general location of the turbidity maximum is a zone of muddy bed sediment containing a high organic content. This 'mud reach' forms a zone of shoaling which often causes navigational problems requiring dredging.

There are three processes that contribute to generating and sustaining the turbidity maximum: vertical gravitational circulation, tidal pumping and sediment dynamics (Officer, 1981).

Vertical Gravitational Circulation

There is a seaward discharge in the riverine section landward of the null point, but a two-layer flow seaward of that point with a landward directed mean flow. Suspended sediment brought down the river will meet the head of the salt intrusion. Landward of the intrusion the high tidal velocities will keep some of the sediment in suspension, whereas the fraction with the higher settling velocities in the lower part of the flow will tend to deposit. The suspended material will be carried downstream by the residual flow in the surface layer until it reaches the middle part of the estuary where the reduced vertical stratification and the increased vertical mixing, together with settling, allow exchange of the particles into the lower layer. There the sediment will be carried towards the head of the estuary, together with sediment being carried into the estuary from the mouth on the landward flow. At the head of the salt intrusion the tidal currents ensure that the sediment is recycled, and in the process the sediment is sorted according to the settling velocity; the slower settling material is swept down the estuary and the higher settling material tends to be deposited. Geyer (1993) emphasizes the importance of the turbulence in the upper layer near the tip of the salt intrusion. The reduction in turbulence by the stratification allows the sediment to settle and trapping in the lower layer is enhanced.

The maximum concentration is therefore near the null point at the head of the salt intrusion (Figure 9.9). The magnitude of the concentrations has been shown by modelling studies to be dependent on the settling velocities, other things being equal. With slower settling particles the maximum is broader and with higher concentrations.

The residual vertical gravitational circulation produces the broad background turbidity in partially mixed estuaries.

Tidal Pumping

The asymmetry of the tide, as exemplified by the processes described in the Tamar estuary, causes the suspended sediment flux to be greater on the flood tide than on the ebb tide. As we have seen, the tidal asymmetry increases inland until the influence of river flow becomes important in amplifying the ebb tide velocities. This occurs somewhere to landward of the salt intrusion. Thus suspended sediment is pumped towards the head of the estuary throughout the estuary and can bring sediment into the estuary from the mouth. By itself this would not lead to a turbidity maximum because the landward flux would be balanced by a compensating seaward flux on the mean flow, just as it is for salt.

For very fine sediment continually in suspension, the wash load, there should thus be an even distribution along the estuary. However, it will provide a maximum when interacting with the sediment settling and re-entrainment during the tide.

The tidal pumping effect will depend on the variation of the tidal conditions along the estuary, but it will control the location of the maximum, generally causing it to be landward of the tip of the salt intrusion.

Sediment Dynamics

The settling and erosion of sediment causes the sediment particles to become separated from the flow at times during the tidal cycle, and this can be represented by a phase difference between the velocity and the concentration variation. This phase difference can produce a residual flux of sediment even when there is no residual movement of water, providing that the currents are asymmetrical. If the currents are symmetrical no residual flux of sediment would occur without a residual in the water flow.

Lags can be produced by a variety of processes (Dyer, 1995b). *Threshold lag* is produced by the requirement for the velocity to exceed a threshold value to start the sediment moving. In a tidal cycle with a short, fast flood current and a long, slow ebb current, the duration of sediment movement on the ebb tide will decrease much more rapidly than that on the flood as the threshold increases. In the extreme, the movement on the ebb tide may cease altogether. Allen *et al.* (1980) have suggested that this is a major cause of the turbidity maximum in macrotidal estuaries.

Settling lag is the delay caused by the particles settling out of suspension. Before they actually reach the bed they may be advected along on the flow beyond the position where the flow was incapable of maintaining them in suspension. This has been proposed as the mechanism for the landward movement of sediment in shallow tidal areas (van Straaten and Kuenen, 1958). Dronkers (1986) has emphasized the importance of the flow directions in the periods around slack water during which the settling takes place.

Scour lag is the time taken for the particles to disperse throughout the water column after it has been eroded. During this time the movement will be dominated by the flows in the near-bed layer, rather than the depth mean or flows higher up.

Erosion lag relates to the variation of threshold conditions with depth in the sediment created by consolidation. This means that the mass per second per unit of shear stress eroded from the bed at high velocities will be different from that eroded at lower velocities.

Many of these lag processes act simultaneously and their effects will be difficult to separate. Obviously, if the sediment is easily eroded, the resulting concentrations will be higher than if the sediment was hard to erode. Similarly, if the sediment settles rapidly as the currents diminish, then the concentration variations on the tidal time-scale will be high. The sediment dynamics thus controls the magnitude of the peak turbidity maximum and the timing of the peak during the tide.

The fluxes of sediment in the estuary can be quantified in much the same way as the salt fluxes were considered in Chapter 6. The instantaneous flux of suspended sediment through an element of the estuary is

$$F = \int_{0}^{h} uc \, dz$$

Averaging over the tidal cycle using the, now familiar, definitions of the mean, tidal and deviation components, and where depth $= \bar{h} + \boldsymbol{h}$

$$\bar{F} = \bar{h} \langle \bar{u} \rangle \langle \bar{c} \rangle + \overline{\boldsymbol{h} \langle U \rangle} \langle \bar{c} \rangle + \boldsymbol{h} \langle C \rangle \langle \bar{u} \rangle + \bar{h} \, \overline{\langle U \rangle \langle C \rangle} + \overline{\boldsymbol{h} \langle U \rangle \langle C \rangle} + \bar{h} \langle \overline{\bar{u}_{\mathrm{d}} \bar{c}_{\mathrm{d}}} \rangle + \bar{h} \langle \overline{U_{\mathrm{d}} C_{\mathrm{d}}} \rangle$$

In this case $\bar{c}$ and C are the mean and tidal variations in concentration, with the subscript d being the deviations from the cross-sectional means. Angled brackets signifies a mean over the depth.

The first term on the right-hand side is the flux on the non-tidal drift, the Eulerian velocity. The second term is the flux on the Stokes drift. Together these two terms provide the downstream advective sediment flux. The terms 3–5 are the tidal pumping terms that are produced by the phase differences, the lags between the sediment concentration, the velocity and the cross-sectional area. These arise mainly as a consequence of the threshold and erosion lags. Term 6 is the vertical gravitational circulation, arising because of the correlation between the landward bottom mean flow and high near-bed concentrations, and the seawards mean surface flow and lower concentrations. Term 7 arises from the changing forms of the vertical profiles of velocity and concentration during the tide, due mainly to the scour and settling lags.

This approach has been applied to several estuaries by Dyer (1978, 1988b), Su and Wang (1986), and Uncles *et al.* (1984, 1985b) assuming that the estuaries were in steady state. In all cases the tidal pumping terms were at least as important as the gravitational circulation and in many cases they dominated over the residual circulation. In general, it seems that in well-mixed estuaries tidal pumping is more dominant, though in partially mixed estuaries the gravitational circulation does become significant.

In any one estuary it is likely that at the seaward end of the turbidity maximum the peak of sediment concentration will appear close to low slack water because of advection from upstream. The phase relationships then produce an upstream tidal pumping that tends to ensure that material returns landwards and prevents it escaping. Thus there are lower concentrations on the ebb tide than on the flood. At the upper end of the turbidity maximum the reverse happens, with the maximum concentrations occurring at near high water slack. In this region, downstream tidal pumping may occur, or the downstream mean advection may balance weak upstream pumping. In the centre of the turbidity maximum there should be a balance between the downstream mean fluxes, the tidal pumping and the gravitational circulation. Suspension of sediment should be equally effective on flood and ebb tides and the concentrations should be similar.

There are also variations laterally in tidal pumping. Uncles *et al.* (1984) have shown that near the head of the Tamar estuary landward pumping of sediment occurred in the central channel, whereas other sections had weak landward pumping in shallow water and seaward pumping in the channel. However, these differences change with river discharge and tidal range.

River discharge increases will enhance the seaward advection, leading to a net sediment movement down the estuary. Decreases in river discharge will lead to the sum of tidal pumping and gravitational circulation exceeding the downstream advection, and net sediment transport back to the head of the estuary (Uncles *et al.*, 1988).

Both tidal pumping and the circulation will vary with tidal range, leading to a change in trapping during the lunar tide. For instance, the Tay estuary exports sediment to the sea during neap tides, but imports it again on spring tides (Dobereiner and McManus, 1983).

As well as wind, storm discharges of water affect the estuarine circulation and the associated transport of sediment. There have been few studies that adequately cover such events. However, that of Nichols (1977) illustrates the effects of tropical storm Agnes on the Rappabannock River estuary in June 1972. As a result of 30 cm of rain in two days, the initial response was a rapid change from a partly mixed to a salt wedge regime, with the saline limit pushed 20 km downstream and with the maximum suspended sediment concentration above the head of the salt wedge. The strong seaward current in the upper layer carried the sediment seawards and the return current on the bed was weak. After this 'shock' stage there was a 'rebound' period during which decreasing inflow allowed the salt wedge to creep landward, the landward residual flow to strengthen and a turbidity maximum to become re-established. The flood passing down Chesapeake Bay complicated the recovery period, but it was concluded that most of the sediment was trapped within the estuary because the saline intrusion was not pushed clear of the estuary. The residual circulation eventually brought most of the sediment back to the head of the estuary.

However, there are difficulties in establishing clear relationships between the fluxes and the overall sediment balance in an estuary because the flux measurements have large errors and the net fluxes are the small differences between the terms. Nevertheless, it is apparent from studies of sediment mineralogy that marine-derived sediments penetrate to the head of most partially and well-mixed estuaries, where they are mixed with those coming from the river.

The *trapping efficiency* is the total sediment accumulated in the estuary divided by the fluvial input. It can exceed 100% for estuaries where there is significant input from the sea by tidal pumping. For stratified estuaries the trapping efficiency will be low because of the escape of fine sediment in the surface layer.

A *filter efficiency* can also be defined (Schubel and Carter, 1984) as

$$\text{Filter efficiency} = 1 - \frac{\text{Flux of suspended sediment through mouth}}{\text{Flux fresh water} \times \text{mean concentration}}$$

A net import occurs when the filter efficiency $F_E > 1$ and net export when $F_E < 1$. However, it is difficult to measure the mean flux of sediment through an estuary mouth because the errors are larger than the differences between the net flood and ebb fluxes.

Consequently, the trapping of both marine- and river-derived sediment at the head of an estuary will cause it to gradually shorten. It then becomes possible for occasional river floods to carry sediment into the coastal zone beyond the influence of the landward flows which tend to draw it back into the estuary at lower flows. The estuary will then be in a steady-state condition and become an equilibrium estuary, where there will be a balance between the sediment and fluid dynamics.

In the seaward sections of the estuary the flooding tide progresses faster in the channels than over the shallower areas. These shallower areas also are well below the low tide mark so that there is a large cross-sectional area for the water to flow through to supply the volumes required to raise the water level further upstream. The water surface slopes on the ebbing tide are more gentle than on the flood tide and the current velocities are in general lower. As a consequence the channels are flood dominated.

Further up the estuary the shallow parts begin to develop into intertidal areas which are dry at low tide. As a consequence the water depth changes dramatically during the tide. Consideration of the elevation and velocity curves shown in Figure 3.1 for a standing wave indicates that, because the slack water coincides with high tide, there are likely to be equal flood and ebb current strengths and durations during the covering and uncovering of the intertidal flats. However, macrotidal estuaries have a partially progressive tidal wave, the slack water occurs after high water, and the shallow water regions on the sides of the estuary will experience large flood-directed water transports while they are inundated. The shallow water areas are thus flood dominated and the headward flood discharge has to be compensated for by extra ebb discharge. This is concentrated in the deeper channels where the flow is ebb dominated, and the Eulerian velocity there, the non-tidal drift, will be considerably greater than the velocity due to the river discharge.

Of course, the detail will depend on the topography of the estuary and how it interacts with the tidal rise and fall. However, in general, the channels near the mouth will be flood dominated and those near the head will be ebb dominated. Where the change occurs will be in the vicinity of the position where extensive intertidal areas are developed. The channels avoid each other at the location of the change in dominance and tend to interdigitate. The sediment carried upstream in the flood-dominant channels meets that carried seawards in the ebb-dominant channels, and sand banks form. The banks will have a flood-dominant channel on one side and an ebb-dominant channel on the other. The sand tends to circulate around the bank and the bank often migrates in a quasi-cyclical way, the channels shallowing and deepening in turn.

The changes in dominance across the estuary results in the large-scale horizontal mean circulation of water. The dispersion of salt and of contaminants can no longer be considered as a one-dimensional process, but two-dimensional depth-averaged models can be reasonably realistic. For sediment where there are distinct vertical gradients of concentration, even when the salinity is well mixed, three-dimensional models are generally required.

Chapter 10

Flushing and Pollution Distribution Prediction

FLUSHING TIME

It is obvious from what we have already seen of estuaries that increased river flow causes both a downstream movement of the salinity intrusion and a more rapid circulation of water. Thus increased river discharge is accompanied by a more rapid exchange of fresh water with the sea, the volume of fresh water accumulated in the estuary increasing to a lesser extent than does the discharge. The *flushing time* is the time required to replace the existing fresh water in the estuary at a rate equal to the river discharge. This is also known as the *residence time*.

The flushing time $T = V_f/R$ where V_f is the total amount of river water accumulated in the whole or a section of the estuary and R is the river flow. The flushing time changes rapidly with discharge variation at low river flow, but changes slowly at high river flow.

The flushing time can be calculated in several ways.

Tidal Prism Method

The simplest way of considering flushing is that due to the exchange of water by the tide. In this method, the water entering on the flood tide is assumed to become fully mixed with that inside and the volume of sea water and river water introduced equals the volume of the tidal prism, the volume between the high and low tide marks. On the ebb the same volume of water is removed and the fresh water content of it must equal the increment of the river flow. If V is the low tide volume and P the intertidal volume (the tidal prism), then the flushing time in tidal cycles

$$T = \frac{V + P}{P} T$$

where T is the tidal period. In other words the flushing time is the high tide volume divided by the tidal prism volume. Obviously, this approach can also be used for the flushing of water in inlets where there is no river flow.

It has been found that the flushing time calculated this way gives a considerably lower flushing time than calculation using other methods. The exaggerated estimate of the rate of flushing is due to the incomplete mixing of the estuarine water: the fresher water near the head of the estuary cannot reach the mouth during the ebb. Also, some of the water which does escape during the ebb returns on the following flood tide.

Fraction of Fresh Water Method

The mean *fractional fresh water concentration* over any segment is

$$f = \frac{S_s - S_n}{S_s} \tag{10.1}$$

where S_s is the salinity of the undiluted sea water and S_n is the mean salinity in a given segment of the estuary. The total volume of fresh water V_f is found by multiplying the fractional fresh water concentration f by the volume of the estuary segment. Thus

$$T = fV/R \tag{10.2}$$

The cumulative flushing time from the estuary head thus increases towards the estuary mouth and with estuary size, decreases with decreasing river flow and is generally of the order of some tens of days. Some representative values are shown in Table 10.1.

Table 10.1 Flushing times of some estuaries.

Estuary	River flow $(m^3 s^{-1})$	Flushing time (days)
Mersey	26	5.3
Bay of Fundy		76
Severn Estuary	480	100
	100	200
	80	300
Narragansett Bay	250	12
	22	40

Water and Salt Budget Method

This is applicable to estuaries where there is a two-layer exchange with the sea at the mouth. It is an adaptation of Knudsen's hydrographic theorem, but with only a flow of river water through the uppermost section at the head of the estuary. The flushing time then is $T = V(S_2 - S_1)/S_2 R$, where S_2 and S_1 are the salinity of the lower and upper layers, respectively, at the mouth.

Modified Tidal Prism Method

Ketchum (1951) modified the tidal prism approach by dividing the estuary into segments, the lengths of which are determined by the excursion of a water particle during the tide, rather than being arbitrary. The segmentation assumes that on the flood tide the low tide volume of a segment completely displaces the water from the next segment landwards, the low tide volume becoming the high tide volume of that segment. This concept has been further developed by Dyer and Taylor (1973) and Wood (1979) and forms a useful introduction to the essentials of estuarine modelling.

The innermost section is that above which the intertidal volume P_0 is supplied by the river flow during the flooding tide (Figure 10.1). Thus $P_0 = R/2$. The section 0 is thus the limit where there is no flow on the flood tide. The limit of the next segment is placed so that the low tide volume equals the high tide volume of the segment 0, i.e. P_0 plus the river flow during the ebb tide. Thus $V_1 = R$. The tidal prism in segment 1 is then defined as being a proportion α of the low tide volume of segment 2. Thus $\alpha V_2 = P_1$. The parameter α allows for incomplete mixing at high water and the fact that not all water can exchange between segments on each tide. Subsequently, in general

$$\alpha V_{n+1} = \alpha V_n + P_n$$

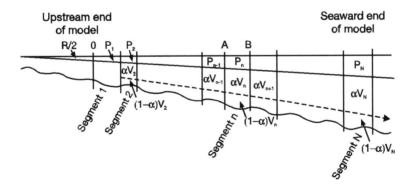

Figure 10.1 Tidal prism segmentation of an estuary. Reproduced by permission of Academic Press from Dyer and Taylor (1973).

Each segment contains, at high tide, a proportion of the volume of water contained in the next seaward segment at low tide, together with some residual water not transferred to the next landward segment. Thus the limits of the segments are equal to the average excursion of a particle of water on the flood tide.

High and low water concentrations of river water C_n^H, C_n^L are defined for each segment. We now consider mixing at high and low water and the requirement that there has to be a seaward flux of river water equal to RC_0 per tidal cycle.

At high water

$$(V_n + P_n)C_n^H = \alpha V_{n+1}C_{n+1}^L + (1 - \alpha)V_n C_n^L; \qquad n \geqslant 2 \qquad (10.3)$$

At low water

$$V_n C_n^L = (\alpha V_n + R)C_{n-1}^H + [(1 - \alpha)V_n - R]C_n^H; \qquad n \geqslant 2 \qquad (10.4)$$

We can assume that $C_{n+1}^L = 0$ at the mouth, where the pure sea water enters, and the equations can be sequentially computed to obtain the high and low water distributions of fresh water (or salinity) in the estuary. Dyer and Taylor (1973) give examples of the results of this for the Raritan and the Thames estuaries. Values of α of 0.5 and 0.9 were obtained.

The flushing time for the segments can be calculated by dividing the volume of fresh water in each segment by the river discharge per tide, and then summation will give the flushing time of the estuary.

Theory of a Mixing Length

Using Ketchum's idea that the element of mixing volume is bounded by the length of the tidal excursion, Arons and Stommel (1951) developed a mixing length theory and attempted to produce flushing numbers to characterize estuaries. If the estuary is of uniform width and of uniform depth (h) and the tide is simultaneous and uniform over the entire channel, then the amplitude of the tidal current U_0 is $U_0 = A_0 \omega(x/h)$, where A_0 is the amplitude of the vertical tidal movement, x is the distance from the head of the estuary and ω is the angular frequency of the tide ($\omega = 2\pi/T$ where T is the tidal period). The amplitude of the horizontal tidal displacement is

$$\xi_0 = -\frac{A_0 x}{h}$$

The horizontal eddy diffusion coefficient K_x is assumed to be related to the tidal displacement and the current by $K_x = 2B\xi_0 U_0$, where B is a constant. Thus

$$K_x = 2BA_0^2 \omega \frac{x^2}{h^2}$$

Introducing a dimensionless length parameter, $\lambda = x/L$ where L is the total estuary length, and a *flushing number*

$$F = \frac{\bar{u}_{\mathrm{f}} h^2}{B A_0^2 \omega L}$$

where u_{f} is the mean velocity of water in the channel due to the river flow, then the one-dimensional diffusion Equation (6.24) becomes

$$F\bar{s} = \lambda^2 \, \mathrm{d}\bar{s}/\mathrm{d}\lambda \tag{10.5}$$

Integrating, $\ln \bar{s} = -F/\lambda + F + \ln S_{\mathrm{s}}$ where S_{s} is the undiluted sea salinity. Thus

$$\bar{s}/S_{\mathrm{s}} = \mathrm{e}^{F(1-1/\lambda)} \tag{10.6}$$

The curves of $\bar{s}/S_{\mathrm{s}}$ against λ for various values of F are shown in Figure 10.2. They show a toe near $\lambda = 0$ and a point of inflexion, similar to the normal picture of salinity distribution. The curves are most sensitive to F in the region $0.1 > F > 10$. Arons and Stommel calculated the values of the constant B by comparison with the observed salinity distributions in the Raritan River and Alberni Inlet. Unfortunately, there was an order of magnitude difference in the

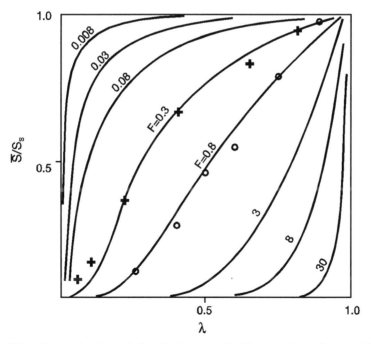

Figure 10.2 Curves showing relationship between flushing number, salinity and length. ○, Raritan River; +, Alberni Inlet. Reproduced from Arons and Stommel *Trans. Am. Geophys. Union*, **32**, 41927, 1951, copyright by the American Geophysical Union.

results, which means that the method fails to predict adequately the intensity of mixing from the given parameters of the estuaries. This may partly be due to the rather unnatural topographic constraints assumed in the analysis, and partly due to the assumed forms of the flushing number and eddy diffusion coefficient.

Flushing Rate

The *flushing rate* is the rate at which the fresh water is exchanged with the sea. Thus $F = V/T = R/f$. The quantity F, which has the dimensions of $m^3 s^{-1}$, represents the combined effects of the tidal exchanges and the gravitational circulation, which together equal the total longitudinal flux in the Hansen–Rattray parameter v (Officer and Kester, 1991).

The tidal exchange flux should be more or less independent of the river discharge, but the gravitational circulation will depend strongly on it. If the tidal exchanges were dominant over the gravitational circulation, then the flushing rate should be about constant for all river discharges. If there were no tides, there would be no flushing when the river flow was zero. In general then, a curve of flushing rate against river discharge should have an intercept value F_{int} at zero river flow, and at finite flows values in excess of this will be the gravitational circulation contribution. The parameter v can then be defined as $v = F_{int}/F$.

Narragansett Bay has a flushing time of about 40 days at low river flow, decreasing to about 10 days at high river flow. Figure 10.3 shows a plot of flushing rate against river flow for the Bay. The dispersive exchange flux F_{int} has a value of about $700 \, m^3 s^{-1}$, and at a river flow of $300 \, m^3 s^{-1}$ the value of v is about 0.28. Thus tidal exchange will exceed $z = 0.50$ at a flushing rate of less than $1400 \, m^3 s^{-1}$, i.e. river flows less than about $130 \, m^3 s^{-1}$, and the contribution of gravitational circulation rises rapidly above that. Consequently, with historical data, it is possible to develop a prediction of flushing and the relative strengths of the gravitational and tidal dispersion characteristics.

Exchange Through the Mouth

We have already alluded to the fact that on the flood tide not all of the water flooding in through the mouth is new sea water: much of it may be water discharged on the ebb tide and mixed with sea water. The volume of fresh water leaving the estuary during the tide must equal the river discharge R, and a summation of the salt flux through the mouth over the ebb tide, assuming steady-state conditions, gives (Van de Kreeke, 1988)

$$R = \varepsilon \, P \, (S_s - S_{out})/S_s \qquad (10.7)$$

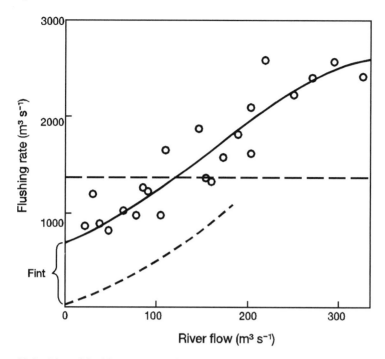

Figure 10.3 Plot of flushing rate F against river flow for Narragansett Bay. Intercept of observations on the F axis gives F_{int}, the diffusive tidal exchange flux. The heavy broken line gives the flushing rate in the absence of gravitational circulation. The thin broken line gives the flushing rate with no tidal exchange, i.e. with only gravitational circulation. Reproduced by permission of Academic Press from Officer and Kester (1991).

where ε is the fraction of ebb water not returning on the flood tide, P is the tidal prism volume, and S_{out} is the discharge weighted average salinity of the ebb water.

Using Equations (10.2) and (10.7) to eliminate the river discharge gives the flushing time T as

$$T = V(S_s - \overline{S}/S_s - S_{out})/\varepsilon P$$

The values of ε will depend on the tidal characteristics and the circulation off the mouth. Typical values are Chincoteague Bay $\varepsilon = 0.34$, San Francisco Bay $\varepsilon = 0.24$, Naruto Strait $\varepsilon = 0.6$, Indian River Bay, Delaware $\varepsilon = 0.23$ and Southampton Water $\varepsilon = 0.71$.

POLLUTION DISPERSION PREDICTION

Near the mouth of the estuary the fresh water fraction is relatively low as the salt water is only slightly diluted. Enough of the mixture must escape on each tide to remove a volume of fresh water equivalent to the river flow. The escaping volume can thus be an order or more greater than the river flow and it is this volume that is available for the dilution and removal of pollutants. Consequently, estuaries are better at diluting and removing pollution than the tributary river. It is obviously useful to be able to predict these effects.

Conservative Pollutants

If a constant rate discharge of a conservative, non-decaying pollutant is made into an estuary the tidal mixing will distribute it both upstream and downstream. The maximum concentration will be in the vicinity of the discharge point. If the pollutant acts in the same way as fresh or salt water, the pollutant distribution will be directly related to the salinity distribution once a steady state has been achieved. Prediction can thus be based on knowledge of the distribution of fresh water in the estuary.

Ketchum (1955) has developed a fractional fresh water method for predicting the concentration of a pollutant. Let the cross-sectional average concentration at the outfall after steady-state conditions have been achieved be C_0. Then

$$C_0 = \frac{P}{R} f_0$$

where P is the rate of supply of the pollutant, R is the river discharge and f_0 the cross-sectional fractional fresh water concentration.

Downstream of the outfall the pollutant must pass through a cross-section at the same rate as it is discharged at the outfall

$$C_x = C_0 \frac{f_x}{f_0} = \frac{P}{R} f_x \tag{10.8}$$

Upstream of the outfall the quantity of pollutant carried upstream with the saline water will balance that carried downstream by the mean flow. Its distribution will be directly proportional to the distribution of salinity, inversely proportional to the fresh water fraction. Thus

$$C_x = C_0 \frac{S_x}{S_0} \tag{10.9}$$

The pollutant distribution will be of the form shown in Figure 10.4. Downstream it will have the same form as the salinity distribution and upstream the inverse of the salinity. It is noticeable that if the discharge point is

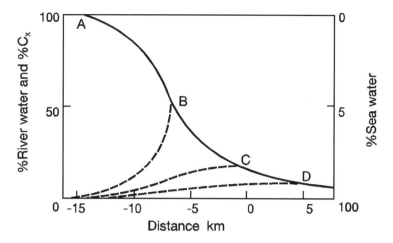

Figure 10.4 Steady-state distribution of a conservative pollutant for four discharge points A, B, C and D. Reproduced by permission of the Water Environment Federation from Ketchum (1952).

moved downstream the concentration levels at points seaward of the outfall are unaffected, but the upstream levels are drastically reduced. The concentration in the immediate vicinity of the outfall is also reduced.

Pritchard (1969) has produced a two-dimensional box model for predicting pollutant distribution which is especially applicable to partially mixed estuaries. He considers the estuary to be divided longitudinally into a number of segments, each of which is divided vertically into two sub-segments. The interface between the vertical segments is the boundary between the seaward flowing surface layer and the landward flowing bottom layer. The continuity of salt and water between each box is then considered assuming negligible longitudinal diffusion.

Referring to Figure 10.5, let $(Q_u)_{n-1,n}$ be the volume flow rate from the $(n-1)$th segment into the nth segment, upper layer and $(Q_L)_{n,n-1}$ the flow in the lower layer from the nth to the $(n-1)$th segment. Similarly, $(Q_u)_{n,n+1}$ will be the upper layer flow rate from the nth to the $(n+1)$th segment, etc. Within the segment n there will be a volume rate of flow due to vertical advection from the lower into the upper layer (Q_v) and a vertical exchange coefficient representing the vertical diffusion, E_n. The salinity of the upper and lower layers is $(S_u)_n$ and $(S_L)_n$, respectively, and the salinity at the boundary between them is $(S_v)_n$. The salinities at the other boundaries will be $(S_u)_{n-1,n}$, $(S_L)_{n+1,n}$, etc. Homogeneity is assumed within each box.

For a steady-state salt distribution, for the upper layer in segment n

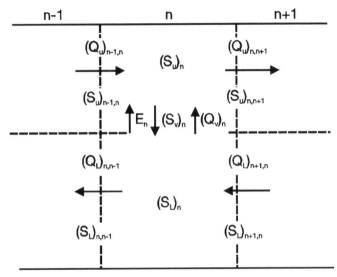

Figure 10.5 Definition diagram for two-dimensional box model. After Pritchard (1969).

$$(S_u)_{n,n+1} \cdot (Q_u)_{n,n+1} = (S_u)_{n-1,n} \cdot (Q_u)_{n-1,n} + \atop E_n[(S_L)_n - (S_u)_n] + (Q_v)_n \cdot (S_v)_n \tag{10.10}$$

From volume continuity

$$(Q_v)_n = (Q_u)_{n,n+1} - (Q_u)_{n-1,n} \tag{10.11}$$

There are similar equations for the lower layer. Similar to Equation (6.20), for any section of the estuary

$$Q_u = R\frac{S_L}{S_L - S_u} \qquad Q_L = R\frac{S_u}{S_L - S_u} \tag{10.12}$$

If the salinity distribution and fresh water flow are known the horizontal volume flow rates can be calculated using Equation (10.12). The vertical flow rates can be calculated from Equation (6.21) and then Equation (10.10) solved for E_n.

Now let a conservative pollutant be introduced into the surface layer of the kth segment at a rate P. The concentration C of the pollutant is assumed uniform in each sub-segment and at the boundaries the concentration is equal to the average value between adjacent segments. Then for the upper layer the pollutant balance is

$$(Q_u)_{n,n+1}\left[\frac{(C_u)_n + (C_u)_{n+1}}{2}\right] = (Q_u)_{n-1,n}\left[\frac{(C_u)_{n-1} + (C_u)_n}{2}\right]$$
$$+ E_n[(C_L)_n - (C_u)_n] + (Q_v)_n\left[\frac{(C_u)_n + (C_L)_n}{2}\right] \tag{10.13}$$

For the lower layer:

$$(Q_L)_{n,n+1}\left[\frac{(C_L)_n + (C_L)_{n+1}}{2}\right] = (Q_L)_{n,n-1}\left[\frac{(C_L)_{n-1} + (C_L)_n}{2}\right]$$
$$+ (Q_v)_n\left[\frac{(C_L)_n + (C_u)_n}{2}\right] + E_n[(C_L)_n - (C_u)_n] \tag{10.14}$$

In the upper layer of the kth segment there is the input term P to be added to Equation (10.13).

Using Equation (10.11), Equation (10.13) and (10.14) become

$$(C_u)_{n-1} \cdot (Q_u)_{n-1,n} - 2(C_u)_n E_n + (C_L)_n[2E_n + (Q_v)_n] - (C_u)_{n+1}(Q_u)_{n,n+1} = 0$$

In the kth segment Equation (10.13) will again have the additional term P.

Using the boundary conditions that upstream C goes to zero and S goes to zero and downstream that $(C_L)_m = 0$ these equations can be solved for the distribution of pollutant concentration for a given input, with the inherent assumption that the vertical exchange coefficient is the same for pollutant as for salt.

An alternative method of calculating the distribution is by use of the one-dimensional diffusion equation (Stommel, 1953b). The net seaward flux of pollutant through any section x is

$$F(x) = Rc - \overline{A}K_x \frac{dc}{dx} \tag{10.15}$$

where R is the river discharge.

Downstream of the source the net flux must be constant and equal to the input. Upstream it will be zero. Providing the diffusion coefficient for the pollutant can be assumed to be the same as that for salt, then we can determine K_x by putting $F(x)$ equal to the river flow and the fresh water fraction f for c. Then

$$K_x = \frac{R(f-1)}{\overline{A}\,df/dx} \tag{10.16}$$

The values of K_x calculated this way are put into Equation (10.15) and the equation is written in a finite difference form and is solved by successive approximation. There are various constraints on the solution: the concentration must approach zero at the ocean and at the head of the estuary and, in the section near the outfall, the difference in the flux upstream and downstream must equal the rate of inflow of pollutant at the source. The methods of Ketchum and Stommel give similar results.

Pollutant dispersion in a one-dimensional estuary has also been considered by Kent (1960). Kent considers a sectionally homogeneous estuary and

$$\frac{\partial \bar{s}}{\partial t} = -\bar{u}\frac{\partial \bar{s}}{\partial x} + \frac{1}{A}\frac{\partial}{\partial x}\left(AK_q\frac{\partial \bar{s}}{\partial x}\right) \tag{10.17}$$

where K_q is the eddy diffusion coefficient for the pollutant.

In the first instance a solution for Equation (10.17) is considered with values of u, A and K_q variable with x. The equation can be written in a finite difference form and, provided the initial distribution of the pollutant in the estuary is known, we can trace the disposition of the pollutant in space and time. This is done by solving the finite difference equation numerically by successive approximations. To make sure the approximation solution converges on the exact one, certain conditions are specified for the increments of time and distance used for the differences.

A solution is also considered with constant coefficients. In this case, Equation (10.17) reduces to

$$\frac{\partial \bar{s}}{\partial t} = -\bar{u}\frac{\partial \bar{s}}{\partial x} + K_q\frac{\partial^2 \bar{s}}{\partial x^2} \tag{10.18}$$

This equation can also be stated in a difference form and solved numerically by successive approximations. Conditions for the space–time increments are again specified to provide convergence of the solution.

In both these analyses it is necessary to know the distribution of the pollutant diffusion coefficient K_q. It is normally considered to be the same as the diffusion coefficient for salt, which can be calculated by Equation (10.16).

Preddy (1954) has modelled pollution in the Thames using a mixing concept not unlike that of Ketchum. He considers that after a tidal cycle a proportion P_1 of the pollutant will be dispersed over a length L downstream and a proportion P_2 over the same length upstream. The amount $1 - P_1 - P_2$ will be at the discharge point. To determine the pollutant distribution it is necessary to determine the values for P_1 and P_2 at different points in the estuary. By considering the continuity of salt and of water, two equations can be formed in which L, P_1 and P_2 are unknowns. The river flow and salinity distribution need to be known and the length L is taken as the tidal excursion. The equations can then be solved by successive approximations. Preddy uses this method to predict the effect on the salinity distribution of changes in river flow.

Non-conservative Pollutants

For non-conservative pollutants prediction becomes more difficult. Concentrations, as well as diminishing because of mixing, also decrease with time and terms have to be included which describe the decay, as well as other processes, such as scavenging of the pollutants on to suspended sediment particles and their deposition. For coliform sewage bacteria the change due to mortality is an exponential effect that can be represented by

$$C_t = C_0\, e^{kt}$$

The constant k is negative and the population decreases to one-tenth in one and a half to three days depending on the light conditions. As this time is comparable with the flushing time, mortality will be important. In the Raritan River practically all of the decrease in bacterial concentration was due to mixing, mortality and grazing by zooplankton. These effects are assumed to be represented by

$$C_n = (C_0)_n\, \frac{r_n}{1-(1-r_n)e^k} \tag{10.19}$$

where r_n is the exchange ratio in segment n.

If the mortality rates and exchange ratios are equal in the segments in the estuary, then downstream of the outfall

$$C_n = (C_0)_n\, \frac{f_n}{f_0} \left(\frac{r}{1-(1-r)e^k} \right)^n \tag{10.20}$$

and upstream

$$C_n = (C_0)_n\, \frac{S_n}{S_0} \left(\frac{r}{1-(1-r)\, e^k} \right)^n \tag{10.21}$$

where n is the number of segments from the outfall, the segments being defined by a modified tidal prism analysis. If the coefficient of mortality is zero, i.e. $e^k = 1$, then these equations equal those for a conservative pollutant. The effect of mortality is shown by

$$\frac{C_n}{C_x} = \left(\frac{r}{1-(1-r)\, e^k} \right)^n \tag{10.22}$$

Larger exchange rates give larger populations in a segment for a given mortality as the water is mixed faster and less decay occurs. Figure 10.6 shows the distribution of non-conservative pollutant in Raritan Bay where the exchange ratio is 0.34. The upper curve shows the effects of dilution alone, with a conservative pollutant discharged from position A. The lower three curves are for the fraction of the population expected when $k = -0.578$ tide^{-1}. Multiplication of the value in the upper curve by that in the lower gives the expected concentration relative to a population size of unity for the pollutant

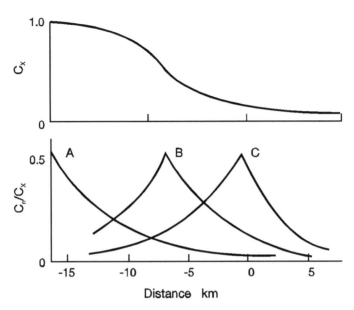

Figure 10.6 Distribution of a non-conservative pollutant for three discharge points in Raritan Bay. For explanation, see text. Reproduced by permission of the Water Environment Federation from Ketchum (1952).

mixed in water with no mortality. This example gives relative populations at the outfall locations of 0.538 for A, 0.278 for B and 0.092 for C.

The peak concentration is thus decreased by the downstream movement of the outfall, but to a greater extent than for a conservative pollutant. In contrast with a conservative pollutant, the concentration at a point downstream of the outfall is reduced if the outfall is moved upstream.

Stommel (1953b) has also introduced a term into Equation (10.15) to allow for decay of pollutant. The equation becomes

$$\frac{\mathrm{d}}{\mathrm{d}x}\left(Rc - \overline{A}K_x\frac{\mathrm{d}c}{\mathrm{d}x}\right) + \frac{\overline{A}c}{t} = 0 \tag{10.23}$$

where t is the time for the concentration to decay to $1/e$ of its initial concentration. This equation can be solved by the same successive approximation method as that used for conservative pollutants.

Many other methods for determining pollutant distributions in estuaries have been published based on the advection–dispersion equation with terms for sources and sinks. The consideration of flushing time and the methods discussed for pollution distribution prediction are mathematical models of increasing complexity. Because they use fresh or salt water as the tracer the results will only apply to substances that act as and are introduced in the same way as either of these. The methods rely on using the existing salinity distribution and the principles of salt continuity to determine the longitudinal dispersion coefficients or exchange coefficients which are then used for the pollutant distribution.

The most commonly modelled non-conservative pollutant is dissolved oxygen. Though it is not a pollutant, it is a mirror for pollution in which the biological demand for oxygen is large. Consequently, a sewage disposal into an estuary will deplete the oxygen levels and to a large extent the dissolved oxygen distribution will be the inverse of the effluent distribution. Modelling the dissolved oxygen content is important so that conditions leading to low oxygen levels and possible anaerobic bacterial activity can be avoided.

The advection–dispersion equation is a mathematical model that is suitable for analysis using computers which can quickly do the large amount of tedious calculation necessary in the successive approximations of the finite difference techniques. However, even then, to make the problem tractable within reasonable computation time, it is necessary to reduce the estuary to a one- or two-dimensional problem. Three-dimensional models are now increasingly used as computer speed and capacity increase, but these models require large quantities of good quality real data taken simultaneously at many stations over several tidal cycles to determine the initial boundary conditions and for validation. Thus the exchange coefficients are determined empirically rather than analytically. Often the amount of real data for validation is limited and it is difficult to assess the accuracy of predictions. The techniques of modelling are well described by Dyke (1996).

Glossary

Advection   The transport due to the mean flow.
Arrested salt wedge   A salt intrusion where the length is controlled by a balance between horizontal pressure gradients and the friction on the interface.
Axial convergence   The flow of water on the surface of an estuary towards the middle from both sides.

Baroclinic flows   Flows induced by horizontal pressure forces resulting from density differences.
Barotropic flows   Flows induced by the horizontal forces resulting from a slope in the water surface.
Barotropic Froude number   The ratio of the tidal amplitude to the water depth.

Centrifugal force   A sideways force acting on a fluid particle because of its movement in a curve.
Circulation parameter   The ratio of the surface residual current to the depth or cross sectional mean current.
Coefficient of longitudinal dispersion   A coefficient in an equation relating the cross-sectionally averaged flux of salt to the longitudinal gradient of salinity.
Coriolis force   The sideways force on an object or particle moving with respect to the earth. It acts to the right in the northern hemisphere.

Densimetric Froude number   The ratio of the current speed to the velocity of a long wave on the density interface.
Destratification events   The abrupt decrease in stratification following mixing due to an internal hydraulic jump.
Diffusive fraction   The fraction of the salt balance that is carried by processes relating to the tidal oscillatory currents.
Dispersion   The spreading of a scalar, such as salt, due to velocity shear.

Ebb and flood channels   Channels where either ebb or flood dominance may occur.
Ebb dominance   A tidal oscillation in velocity that has ebb currents greater than flood currents.
Eddy diffusivity (eddy diffusion coefficient)   A coefficient relating the turbulent diffusion of a variable to its gradient.
Eddy viscosity   A coefficient relating the turbulent shear stress to the velocity gradient.

Ekman transport   The transport of water at right angles to the wind direction.
Entrainment   The process of an upward transport of salt by waves breaking on the halocline.
Entrainment velocity   The upward vertical velocity involved in entrainment.
Entropy   Where there is a constant relationship between the work the tidal wave is doing per unit area of the sea bed and the energy expended in moving sediment.
Equation of state   Equation relating the density of sea water to the salinity.
Equilibrium estuary   An estuary where there is no net sedimentation or erosion despite considerable sediment transport.
Erosion lag   A phase lag between the current velocity and the suspended sediment concentration due to consolidation effects in the bed.
Estuarine Froude number   A Froude number formed from the river discharge and overall estuary characteristics.
Estuarine Richardson number   The ratio of the gain in potential energy due to the river discharge to the mixing energy of the tidal currents.
Estuary number   As defined in Equation (2.1).
Eulerian velocity   A velocity measured with time at a fixed point in space.
Eustatic   Variations of sea level due to changes in the volume of sea water.

Fick's equation   An equation which states that the advection of salt is balanced by diffusion and by any temporal change in concentration.
Filter efficiency   The efficiency of an estuary in discharging through the mouth the suspended sediment brought in by the river.
Fjord salinity loop   A time change of the relationship between the salinity difference between the mouth and head of a fjord and the vertical salinity difference at the mouth.
Flood dominance   A tidal oscillation in velocity that has flood currents greater than ebb currents.
Flow ratio   The ratio of river flow per tidal cycle to the tidal prism.
Flushing rate   The rate at which fresh water is exchanged with the sea through the mouth.
Flushing time   The time required to replace the fresh water in an estuary at a rate equal to the river discharge.
Flux Richardson number   The fraction of turbulent energy used in vertical mixing.
Fractional fresh water concentration   The proportion of the volume of an estuarine segment occupied by fresh water.
Froude number   A dimensionless number comparing the speed of the flow to the velocity of propagation of a surface long wave.

Gradient Richardson number   A dimensionless Richardson number formed from the vertical gradients of density and velocity.

Halocline   The interface between layers of different salinity.
Holmboe waves   Sharp-crested interfacial waves involved in the process of entrainment.
Hydraulic jump   A drastic change of flow velocity and water depth when the Froude number reaches unity.
Hydrostatic equation   An equation expressing the vertical pressure forces in balance with the gravitational acceleration.
Hypersynchronous estuaries   Estuaries where convergence exceeds friction and the tidal range increases towards the head.
Hypertidal   A tidal range greater than 6 m.
Hyposynchronous estuaries   Where friction exceeds the effects of convergence and the tidal range decreases towards the head.

Intense mixing periods Times when intense mixing takes place due to critical internal Froude number conditions.
Internal Froude number A Froude number formed from the layer thicknesses and densities.
Internal waves Waves formed on the density interface between two layers of water.
Isostatic Variations of sea level due to changing elevation of the land.

Kelvin–Helmholtz instabilities Waves created by shear across the interface between two moving layers that break down to turbulent mixing.
Kinetic energy The energy that an object or particle has because of its motion.

Lagrangian currents Currents measured by following particles moving with the flow.
Layer Richardson number A Richardson number formed from the relative densities and velocities of a layered flow.
Lee waves Stationary internal waves created in stratified flow over topographic changes of water depth.
Level of no motion The depth at which the tidally averaged or residual flow reverses its direction.

Macrotidal A tidal range between 4 and 6 m.
Mesotidal A tidal range between 2 and 4 m.
Microtidal A tidal range less than 2 m.

Newton's second law of motion An equation that states that force equals mass times acceleration.
Non-tidal drift The tidally averaged Eulerian velocity.
Null point The position in an estuary where the level of no motion reaches the bed.

O'Brien relationship An empirical relation between the cross-sectional area at the mouth and the tidal prism volume.
Overmixed The limit set on the occurrence of two-way flow by the critical densimetric Froude number.
Overstraining The diminution of stratification by tidal straining and the restoration of homogeneous conditions.

Potential energy The energy that an object or particle has because of its elevation or distance from the earth.
Prandtl number The ratio of eddy diffusivity to eddy viscosity.
Progressive wave A wave where the maximum horizontal velocity is at the crest and the trough of the wave and the crest advances in time and space.
Pycnocline The interface between two layers of different density.

Quadratic friction law A law relating the bed shear stress to the square of the flow velocity.

Residence time The time taken to replace the fresh water in an estuary at a rate equal to the river discharge. Also known as the flushing time.
Reynolds flux The turbulent flux of a scalar property such as salt.
Reynolds number A dimensionless number relating the effects of viscosity to those of turbulence.
Reynolds stress The turbulent flux of momentum, the frictional force due to turbulence.
Richardson number A dimensionless number relating the stabilizing effects of vertical density stratification to the destabilizing effects of velocity shearing.

Salt continuity   An equation which states that salt is not created or destroyed.

Scour lag   A phase lag between the current velocity and the suspended sediment concentration due to the time taken for eroded sediment to disperse through the water column.

Secondary flow   Components of the flow at right angles to the axis of the channel.

Settling lag   A phase lag between the current velocity and the suspended sediment concentration due to the settling of particles through the water towards the bed.

Standing wave   A wave formed by reflection of a progressive wave or superimposition of two opposite travelling progressive waves. The maximum horizontal velocity occurs at mean water level.

Steady flow   A flow that does not vary in velocity with time.

Stokes drift   The forward motion of a particle due to the interaction of a wave with the changing water depth.

Stratification number   As defined in Equation (2.4).

Stratification parameter   The ratio of the surface to bottom salinity difference to the depth or cross-sectional mean salinity.

Synchronous estuaries   Estuaries where convergence and friction are equal and the tidal range is constant along the estuary.

Thermohaline convection   The sinking of cold dense water and the rising of warmer less dense water.

Threshold lag   A phase lag between the current velocity and the suspended sediment concentration due to the threshold of sediment movement.

Tidal intrusion front   A convergence near the mouth of an estuary or fjord where the inflowing saline water plunges beneath the impounded fresher water.

Tidal prism   The volume of water within the estuary between high and low tides.

Tidal pumping   The tidally averaged transport of salt due to phase differences in the velocity, the salinity and the water depth.

Tidal straining   The differential movement of two layers of water due to tidal friction at the base of the lower layer slowing it down relative to the upper layer.

Trapping efficiency   The total sediment accumulated in an estuary divided by the fluvial input.

Turbulent diffusion   The action of turbulence in mixing water with different characteristics and distributing those characteristics more evenly.

Uniform flow   A flow that does not vary in strength in spatial dimensions.

Vertical gravitational circulation   The density induced flow in which a saline layer tends to flow underneath a less saline layer which tends to flow in the opposite direction, with mixing between them.

Volume continuity   Equation expressing the fact that volume is preserved in flowing water.

von Karman–Prandtl equation   An equation formalizing the logarithmic velocity profile near the bed.

References

Abraham G, 1980. On internally generated estuarine turbulence. In: *Second International Symposium on Stratified Flows, AIHR* (Eds T Carstens and T McClimans), Tapir, Trondheim, 344–353.

Allen GP, Salomon JC, Bassoulet P, Du Penhoat Y and De Grandpre C, 1980. Effects of tides on mixing and suspended sediment transport in macrotidal estuaries. *Sedim. Geol.* **26**, 69–90.

Anderson JJ and Devol AH, 1973. Deep water renewal in Saanich Inlet, an intermittently anoxic basin. *Est. Coast. Mar. Sci.* **1**, 1–10.

Anwar HO, 1983. Turbulence measurements in stratified and well-mixed estuarine flows. *Est. Coast. Shelf Sci.* **17**, 243–260.

Armi L and Farmer DM, 1986. Maximal two-layer exchange through a contraction with barotropic net flow. *J. Fluid Mech.* **164**, 27–51.

Arons AB and Stommel H, 1951. A mixing length theory of tidal flushing. *Trans. Am. Geophys. Union* **32**, 41 921.

Blanton J, 1969. Energy dissipation in a tidal estuary, *J. Geophys. Res.* **74**, 5460–5466.

Boon JDIII and Byrne RJ, 1981. On basin hypsometry and the morphodynamic response of coastal inlet systems. *Mar. Geol.* **40**, 27–48.

Bowden KF, 1960. Circulation and mixing in the Mersey Estuary, *IASH Comm. Surface Waters, Publ. 51*, 352–360.

Bowden KF, 1963. The mixing processes in a tidal estuary. *Int. J. Air Wat. Pollut.* **7**, 343–356.

Bowden KF, 1965. Horizontal mixing in the sea due to a shearing current. *J. Fluid Mech.* **21**, 83–95.

Bowden KF, 1981. Turbulent mixing in estuaries. *Ocean Mangm.* **6**, 117–135.

Bowden KF and Gilligan RM, 1971. Characteristic features of estuarine circulation as represented in the Mersey Estuary. *Limnol. Oceanogr.* **16**, 490–502.

Bowden KF and Sharaf el Din SH, 1966a. Circulation, salinity and river discharge in the Mersey Estuary. *Geophys. J. Roy. Astron. Soc.* **10**, 383–400.

Bowden KF and Sharaf el Din SH, 1966b. Circulation and mixing processes in the Liverpool Bay area of the Irish Sea. *Geophys. J. Roy. Astron. Soc.* **11**, 279–292.

Bowden KF, Fairbairn LA and Hughes P, 1959. The distribution of shearing stresses in a tidal current. *Geophys. J. Roy. Astron. Soc.* **2**, 288–305.

Bowman MJ, 1988. Estuarine fronts. In: *Hydrodynamics of Estuaries* (Ed. B Kjerfve), CRC Press, Boca Raton, 85–132.

Brown WS and Trask RP, 1980. A study of tidal energy dissipation and bottom stress in an estuary. *J. Phys. Oceanogr.* **10**, 1742–1754.

Cameron WM, 1951. On the transverse forces in a British Columbia Inlet. *Trans. Roy. Soc. Can.* **45**, 1–9.

Cameron WM and Pritchard DW, 1963. Estuaries. In: *The Sea* (Ed. MN Hill), Vol. 2, Wiley, New York, 306–324.

Cannon GA and Laird NP, 1978. Variability of currents and water properties from year-long observations in a fjord estuary. In: *Hydrodynamics of Estuaries and Fjords* (Ed. JCJ Nihoul), Elsevier Oceanography Series, Elsevier, Amsterdam, 515–535.

Dalrymple RW, Zaitlin BA and Boyd R, 1992. A conceptual model of estuarine sedimentation. *J. Sedim. Petrol.* **62**, 1130–1146.

Davies JH, 1964. A morphogenetic approach to world shorelines. *Z. Geomorphol.* **8**, 127–142.

Dionne JC, 1963. Towards a more adequate definition of the St. Lawrence estuary. *Z. Geomorphol.* **7**, 36–44.

Dobereiner C and McManus J, 1983. Turbidity maximum migration and harbour siltation in the Tay Estuary. *Can. J. Fish. Aquat. Sci.* **40**, suppl. 1, 117–129.

Doyle BE and Wilson RE, 1978. Lateral dynamic balance in the Sandy Hook to Rockaway Point transect. *Est. Coast. Mar. Sci.* **6**, 165–174.

Dronkers J, 1986. Tidal asymmetry and estuarine morphology. *Neth. J. Sea Res.* **20**, 117–131.

Dyer KR, 1974. The salt balance in stratified estuaries. *Est. Coast. Mar. Sci.* **2**, 273–281.

Dyer KR, 1977. Lateral circulation effects in estuaries. In: *Estuaries, Geophysics, and the Environment*, National Academy of Sciences. Washington DC, 22–29.

Dyer KR, 1978. The balance of suspended sediment in the Gironde and Thames estuaries. In: *Estuarine Transport Processes* (Ed. BJ Kjerfve), Univ. South Carolina Press, 135–145.

Dyer K, 1983. Mixing caused by lateral internal seiching within a partially mixed estuary. *Est. Coast. Shelf Sci.* **26**, 51–66.

Dyer KR, 1986. *Estuarine and Coastal Sediment Dynamics*. Wiley, Chichester, 342 pp.

Dyer KR, 1988a. Tidally generated estuarine mixing processes. In: *Hydrodynamics of Estuaries* (Ed. BJ Kjerfve), Vol. 1, CRC Press, Boca Raton, 41–57.

Dyer KR, 1988b. Fine sediment particle transport in estuaries. In: *Physical Processes in Estuaries* (Eds J Dronkers and W van Leussen), Springer, Berlin, 295–310.

Dyer KR, 1989. Estuarine flow interaction with topography: lateral and longitudinal effects. In: *Estuarine Circulation* (Eds BJ Neilson, A Kuo and J Brubaker), Humana Press, Clifton NJ, 39–59.

Dyer KR, 1991. Circulation and mixing in stratified estuaries. *Mar. Chem.* **32**, 111–120.

Dyer KR, 1995a. Response of estuaries to climate change. In: *Climate Change Impact on Coastal Habitation* (Ed. D Eisma), Lewis, Boca Raton, 85–110.

Dyer KR, 1995b. Sediment transport processes in estuaries. In: *Geomorphology and Sedimentology of Estuaries* (Ed. GME Perillo), Elsevier Science, Oxford, 423–449.

Dyer KR and New AL, 1986. Intermittency in estuarine mixing. In: *Estuarine Variability* (Ed. DA Wolfe), Academic Press, Orlando, 321–339.

Dyer KR and Ramamoorthy K, 1969. Salinity and water circulation in the Vellar Estuary. *Limnol. Oceanog.* **14**, 4–5.

Dyer KR and Taylor PA, 1973. A simple, segmented prism model of tidal mixing in well-mixed estuaries. *Est. Coast. Mar. Sci.* **1**, 411–418.

Dyer KR, Gong WK and Ong JE, 1992. The cross sectional salt balance in a tropical estuary during a lunar tide and a discharge event. *Est. Coast. Shelf Sci.* **34**, 579–591.

Dyke PPG, 1996. *Modelling Marine Processes*, Prentice-Hall, London, 152 pp.

Ebbesmeyer CC, Barnes CA and Langley CW, 1975. Applications of an advective–diffusion equation to a water parcel observed in a fjord. *Est. Coast. Mar. Sci.* **3**, 249–268.

Elder JW, 1959. The dispersion of marked fluid in turbulent shear flow. *J. Fluid Mech.* **5**, 544–560.

Elliott AJ, 1978. Observations of the meteorologically induced circulation in the Potomac Estuary. *Est. Coast. Mar. Sci.* **6**, 285–299.

Elliott AJ and Wang DP, 1978. The effect of meteorological forcing on Chesapeake Bay: the coupling between an estuary system and its adjacent waters. In: *Hydrodynamics of Estuaries and Fjords* (Ed. JCJ Nihoul), Elsevier Oceanography Series No. 23, 127–145.

Fairbridge RW, 1980. The estuary: its definition and geodynamic cycle. In: *Chemistry and Biogeochemistry of Estuaries* (Eds E Olausson and I Cato), Wiley, New York, 1–35.

Farmer DM and Armi L, 1986. Maximal two-layer exchange over a sill and through the combination of a sill and contraction with barotropic flow. *J. Fluid Mech.* **164**, 53–76.

Farmer DM and Freeland HJ, 1983. The physical oceanography of fjords. *Prog. Oceanogr.* **12**, 147–219.

Farmer DM and Osborn TR, 1976. The influence of wind on the surface layer of a stratified inlet. *J. Phys. Oceanogr.* **6**, 931–940.

Farmer HG and Morgan GW, 1953. The salt wedge. In: *Proceedings of the 3rd Conference on Coastal Engineering*, ASCE, Washington DC, 54–64.

Farrell SC, 1970. *Sediment Distribution and Hydrodynamics Saco River and Scarboro Estuaries, Maine*. Cont. 6-CRG, Dept. Geol., Univ. Mass.

Fernando HJS, 1991. Turbulent mixing in stratified fluids. *Annu. Rev. Fluid Mech.* **23**, 455–493.

Fischer HB, 1972. Mass transport mechanisms in partially stratified estuaries. *J. Fluid Mech.* **53**, 672–687.

Fischer HB, 1973. Longitudinal dispersion and turbulent mixing in open-channel flow. *Annu. Rev. Fluid Mech.* **5**, 59–78.

Fischer HB, 1976. Mixing and dispersion in estuaries. *Annu. Rev. Fluid Mech.* **8**, 107–133.

Fitzgerald DM and Nummedal D, 1983. Response characteristics of an ebb-dominated tidal inlet channel. *J. Sedim. Petrol.* **53**, 833–845.

Friedrichs CT and Aubrey DG, 1988. Non-linear tidal distortion in shallow well-mixed estuaries: a synthesis. *Est. Coast. Shelf Sci.* **27**, 521–545.

Gao S and Collins M, 1994. Tidal inlet equilibrium, in relation to cross-sectional area and sediment transport patterns. *Est. Coast. Shelf Sci.* **38**, 157–172.

Gardner GB and Smith JD, 1978. Turbulent mixing in a salt wedge estuary. In: *Hydrodynamics of Estuaries and Fjords* (Ed. JCJ Nihoul), Elsevier, Amsterdam, 79–106.

Garvine RW, 1977. River plumes and estuary fronts. In: *Estuaries, Geophysics and the Environment*, National Academy of Sciences, Washington DC, 30–35.

Garvine RW, 1985. A simple model of estuarine subtidal fluctuations forced by local and remote wind stress. *J. Geophys. Res.* **90**, 11 945–11 948.

Geyer WR, 1988. The advance of a salt wedge: observations and dynamic model. In: *Physical Processes in Estuaries* (Ed. J Dronkers and W van Leussen), Springer-Verlag, Berlin, 181–195.

Geyer WR, 1993. The importance of suppression of turbulence by stratification on the estuarine turbidity maximum. *Estuaries* **16**, 113–125.

Geyer WR and Farmer DM, 1989. Tide-induced variation of the dynamics of a salt wedge estuary. *J. Phys. Oceanogr.* **19**, 1060–1072.

Geyer WR and Smith JD, 1987. Shear instability in a highly stratified estuary. *J. Phys. Oceanogr.* **17**, 1668–1679.

Guymer I and West JR, 1988. The determination of estuarine diffusion coefficients using a fluorometric dye tracing technique. *Est. Coast. Shelf Sci.* **27**, 297–310.

Guymer I and West JR, 1991. Field studies of the flow structure in a straight reach of the Conwy Estuary. *Est. Coast. Shelf Sci.* **32**, 581–596.

Hansen DV, 1965. Currents and mixing in the Columbia River estuary. In: *Transactions of the Joint Conference on Ocean Science and Ocean Engineering*, Marine Technical Society and American Society of Limnology and Oceanography, Washington DC, 943–955.

Hansen DV and Rattray M Jr, 1966. New dimensions in estuary classification. *Limnol. Oceanog.* **11**, 319–326.

Harleman DRF and Ippen AT, 1960. The turbulent diffusion and convection of saline water in an idealized estuary. *IASH Comm. Surface Water, Publ. 51*, 362–378.

Hass LW, 1977. The effect of the spring–neap tidal cycle on the vertical salinity structure of the James, York and Rappahannock Rivers, Virginia. *Est. Coast. Mar. Sci.* **5**, 485–496.

Heathershaw AD, 1974. 'Bursting' phenomena in the sea. *Nature* **296**, 394–395.

Hughes P, 1958. Tidal mixing in the Narrows of the Mersey Estuary. *Geophys. J. Roy. Astron. Soc.* **1**, 271–283.

Hughes RP and Rattray M Jr, 1980. Salt flux and mixing in the Columbia River estuary. *Est. Coast. Mar. Sci.* **10**, 479–494.

Huzzey LM and Brubaker JM, 1988. The formation of longitudinal fronts in a coastal plain estuary. *J. Geophys. Res.* **93**, 1329–1334.

Ippen AT, 1966. *Estuary and Coastline Hydrodynamics*, McGraw-Hill, New York.

Ippen AT and Harleman DRF, 1961. One-dimensional analysis of salinity intrusion in estuaries. *Tech. Bull. 5, Comm. Tidal Hydraul. Corps. Eng. US Army.*

Jackson RG, 1976. Sedimentological and fluid-dynamic implications of the turbulent bursting process in geophysical flows. *J. Fluid Mech.* **77**, 531–560.

Jay DA, 1990. Residual circulation in shallow estuaries: shear, stratification and transport processes. In: *Residual Currents and Long-term Transport* (Ed. RT Cheng), *Coastal & Estuarine Studies* **38**, Springer-Verlag, New York, 49–63.

Jay DA and Musiak JD, 1996. Internal tidal asymmetry in channel flows: origins and consequences. In: *Mixing Processes in Estuaries and Coastal Seas* (Ed. C Pattiaratchi), *Coastal and Estuarine Studies Series* **50**. American Geophysical Union, Washington, 219–258.

Jay DA and Smith JD, 1988. Residual circulation in and classification of shallow, stratified estuaries. In: *Physical Processes in Estuaries* (Eds J Dronkers and W van Leussen), Springer-Verlag, Berlin, 21–41.

Jay DA and Smith JD, 1990a. Residual circulation in shallow estuaries 1. Highly stratified, narrow estuaries. *J. Geophys. Res.* **95**, 711–731.

Jay DA and Smith JD, 1990b. Residual circulation in shallow estuaries 2. Weakly stratified and partially mixed, narrow estuaries. *J. Geophys Res.* **95**, 733–748.

Kent RE, 1960. Diffusion in a sectionally homogeneous estuary. *Proc. Am. Soc. Civil Eng.* **86**, SA 2, 15–47.

Ketchum BH, 1951. The exchanges of fresh and salt water in tidal estuaries. *J. Mar. Res.* **10**, 18–38.

Ketchum BH, 1952. Circulation in estuaries. *Proc. 3rd Conference Coastal Engineering.* Amer. Soc. Civil Eng., Washington, 65–76.

Ketchum BH, 1955. Distribution of coliform bacteria and other pollutants in tidal estuaries. *Sewage Indust. Wastes* **27**, 1288–1296.

Kjerfve B, 1979. Measurement and analysis of water current, temperature, salinity, and density. In: *Estuarine Hydrography and Sedimentation* (Ed. KR Dyer), Cambridge University Press, Cambridge, 230 pp.

Kjerfve B (Ed.), 1988. *Hydrodynamics of Estuaries*, Vols 1 & 2. CRC Press, Boca Raton.

Kjerfve B, Stevenson LH, Proehl JA, Chrzanowski TH and Kitchens WM, 1981. Estimation of material fluxes in an estuarine cross section: a critical analysis of spatial measurement density and errors. *Limnol. Oceanogr.* **26**, 325–335.

Knight DW, 1981. Some field measurements concerned with the behaviour of resistance coefficients in a tidal channel. *Est. Coast. Shelf Sci.* **12**, 303–322.

Kullenberg G, 1977. Entrainment velocity in natural stratified vertical shear flow. *Est. Coast. Mar. Sci.* **5**, 329–338.

Largier JL, 1992. Tidal intrusion fronts. *Estuaries* **15**, 26–39.

Largier JL and Taljaard S, 1991. The dynamics of tidal intrusion, retention, and removal of seawater in a bar-built estuary. *Est. Coast. Shelf Sci.* **33**, 325–338.

Lauff GH, 1967. *Estuaries*, Publ. 83, American Association for the Advancement of Science.

Lewis RE, 1979. Transverse velocity and salinity variations in the Tees Estuary. *Est. Coast. Mar. Sci.* **8**, 317–326.

Lewis RE, 1996. Relative contributions of interfacial and bed generated mixing in the estuarine energy balance. In: *Mixing in Estuaries and Coastal Seas* (Ed. C Pattiaratchi), *Coastal and Estuarine Studies Series* **50**. American Geophysical Union, Washington DC, 250–266.

Lewis RE and Lewis JO, 1983. The principal factors contributing to the flux of salt in a narrow, partially stratified estuary. *Est. Coast. Mar. Sci.* **16**, 599–626.

Linden PF, 1979. Mixing in stratified fluids. *Geophys. Astrophys. Fluid Dyn.* **13**, 2–23.

McEwan AD, 1983. The kinematics of stratified mixing through internal wave breaking. *J. Fluid Mech.* **128**, 47–57.

Munk WH and Anderson ER, 1948. Notes on a theory of the thermocline. *J. Mar. Res.* **7**, 276–295.

Murray SP and Siripong A, 1978. Role of lateral gradients and longitudinal dispersion in the salt balance of a shallow, well-mixed estuary. In: *Estuarine Transport Processes* (Ed. B Kjerfve), University South Carolina, 113–124.

Nepf HM and Geyer WR, 1996. Intra-tidal variations in stratification and mixing in the Hudson Estuary. *J. Geophys. Res.* **101**, 12 079–12 086.

New AL, Dyer KR and Lewis RE, 1986. Predictions of the generation and propagation of internal waves and mixing in a partially stratified estuary. *Est. Coast. Shelf Sci.* **22**, 199–214.

New AL, Dyer KR and Lewis RE, 1987. Internal waves and intense mixing periods in a partially stratified estuary. *Est. Coast. Shelf Sci.* **24**, 15–33.

Nichols MM, 1977. Response and recovery of an estuary following a river flood. *J. Sedim. Petrol.* **47**, 1171–1186.

Nichols MM and Biggs RB, 1985. Estuaries. In: *Coastal Sedimentary Environments* (Ed. RA Davis), Springer-Verlag, New York, 77–186.

Nunes RA and Simpson JH, 1985. Axial convergence in a well-mixed estuary. *Est. Coast. Shelf Sci.* **20**, 637–649.

Nychas SG, Hershey HC and Brodkey RS, 1973. A visual study of turbulent shear flow. *J. Fluid Mech.* **61**, 513–540.

O'Brien MP, 1969. Equilibrium flow areas of inlets on sandy coasts. *J. Waterways, Harbours Coast. Eng. Div. ASCE* **95**, 43–52.

O'Donnell J., 1993. Surface fronts in estuaries: a review. *Estuaries* **16**, 12–39.

Oey, L-Y, 1984. On steady salinity distribution and circulation in partially mixed and well mixed estuaries. *J. Phys. Oceanogr.* **14**, 629–645.

Offen GR and Kline SJ, 1975. A proposed model of the bursting process in turbulent boundary layers. *J. Fluid Mech.* **70**, 209–228.

Officer CB, 1976. *Physical Oceanography of Estuaries (and associated coastal waters)*. Wiley, New York, 465 pp.

Officer CB, 1977. Longitudinal circulation and mixing relations in estuaries. In: *Estuaries, Geophysics and the Environment*, National Academy of Sciences, Washington DC, 13–21.

Officer CB, 1981. Physical dynamics of estuarine suspended sediment. *Mar. Geol.* **40**, 1–14.

Officer CB and Kester DR, 1991. On estimating the non-advective tidal exchanges and advective gravitational circulation exchanges in an estuary. *Est. Coast. Shelf Sci.* **32**, 99–103.

Okubo A, 1964. Equations describing the diffusion of an introduced pollutant in a one-dimensional estuary. In: *Studies on Oceanography* (Ed. K Yoshida), Univ. Washington Press.

Okubo A, 1973. Effect of shoreline irregularities on streamwise dispersion in estuaries and other embayments. *Neth. J. Sea Res.* **6**, 213–224.

Partch EN and Smith JD, 1978. Time dependent mixing in a salt wedge estuary. *Est. Coast. Mar. Sci.* **6**, 3–19.

Perillo GME (Ed.), 1995. *Geomorphology and Sedimentology of Estuaries*, Elsevier, Amsterdam, 471 pp.

Pickard GL, 1956. Physical features of British Columbia inlets. *Trans. Roy. Soc. Canada*, **50**, 47–58.

Pickard GL, 1961. Oceanographic features of inlets in the British Columbia mainland coast. *J. Fish. Res. Bd. Can.* **18**, 907–999.

Pickard GL, 1971. Some physical oceanographic features of inlets of Chile. *J. Fish. Res. Bd. Can.* **28**, 1077–1106.

Pickard GL and Rodgers K, 1959. Current measurements in Knight Inlet, British Columbia. *J. Fish. Res. Bd. Can.* **16**, 635–678.

Posmentier ES, 1977. The generation of salinity fine structure by vertical diffusion. *J. Phys. Oceanogr.* **7**, 298–300.

Prandle D, 1981. Salinity intrusion in estuaries. *J. Phys. Oceanogr.* **11**, 1311–1324.

Prandle D, 1985. On salinity regimes and the vertical structure of residual flows in narrow tidal estuaries. *Est. Coast. Shelf Sci.* **20**, 615–635.

Prandle D, 1991. Tides in estuaries and embayments [review]. In: *Tidal Hydrodynamics* (Ed. BB Parker), Wiley, New York, 125–152.

Prandle D and Rahman M, 1980. Tidal response in estuaries. *J. Phys. Oceanogr.* **10**, 1552–1573.

Preddy WS, 1954. The mixing and movement of water in the estuary of the Thames. *J. Mar. Biol. Assoc. UK* **33**, 645–662.

Price WA and Kendrick MP, 1963. Field and model investigations into the reason for siltation in the Mersey Estuary. *Proc. Inst. Civil Eng.* **24**, 473–518.

Pritchard DW, 1952a. Estuarine hydrography. *Adv. Geophys.* **1**, 243–280.

Pritchard DW, 1952b. Salinity distribution and circulation in the Chesapeake Bay Estuaries system. *J. Mar. Res.* **11**, 106–123.

Pritchard DW, 1954. A study of the salt balance in a coastal plain estuary. *J. Mar Res.* **13**, 133–144.

Pritchard DW, 1955. Estuarine circulation patterns. *Proc. Am. Soc. Civ. Eng.* **81**, No. 717.

Pritchard DW, 1956. The dynamic structure of a coastal plain estuary. *J. Mar. Res.* **15**, 33–42.

Pritchard DW, 1958. The equations of mass continuity and salt continuity in estuaries. *J. Mar. Res.* **17**, 412–423.

Pritchard DW, 1967. Observations of circulation in coastal plain estuaries. In: *Estuaries* (Ed. GH Lauff), American Association for the Advancement of Science.

Pritchard DW, 1969. Dispersion and flushing of pollutants in estuaries. *Proc. Am. Soc. Civ. Eng.* **95**, HY1, 115–124.

Pritchard DW and Burt WV, 1951. An inexpensive and rapid technique for obtaining current profiles in estuarine waters. *J. Mar. Res.* **10**, 180–189.

Pritchard DW and Kent RE, 1953. The reduction and analysis of data from the James River Operation Oyster Spat. *Tech Rept. VI, Ref. 53–12*, Chesapeake Bay Inst., Johns Hopkins Univ.

Rattray M Jr and Dworski JG, 1980. Comparison of methods for analysis of the transverse and vertical circulation contributions to the longitudinal advective salt flux in estuaries. *Est. Coast. Shelf Sci.* **11**, 515–536.

Rattray M Jr and Mitsuda E, 1974. Theoretical analysis of conditions in a salt wedge. *Est. Coast. Mar. Sci.* **2**, 375–394.

Reeburgh WS, Muench RD and Cooney RT, 1976. Oceanographic conditions during 1973 in Russell Fjord. *Est. Coast. Mar. Sci.* **4**, 129–146.

Reed DJ, 1987. Temporal sampling and discharge asymmetry in salt marsh creeks. *Est. Coast. Shelf Sci.* **25**, 459–466.

Robinson SK, 1991. Coherent motions in the turbulent boundary layer. *Annu. Rev. Fluid Mech.* **23**, 601–639.

Saelen OH, 1967. Some features of the hydrography of Norwegian fjords. In: *Estuaries* (Ed. GH Lauff), American Association for the Advancement of Science.

Schubel JR and Carter HH, 1984. The estuary as a filter for fine-grained sediment. In: *The Estuary as a Filter* (Ed. V Kennedy), Academic Press, New York, 81–105.

Scott CF, 1993. Canonical parameters for estuarine classification. *Est. Coast. Shelf Sci.* **36**, 529–540.

Sharples J, Simpson JH and Brubaker JM, 1994. Observations and modelling of periodic stratification in the Upper York River Estuary, Virginia. *Est. Coast. Shelf Sci.* **38**, 301–312.

Simmons HB, 1955. Some effects of upland discharge on estuarine hydraulics. *Proc. Am. Soc. Civ. Eng.* **81**, No. 792.

Simpson JH and Nunes RA, 1981. The tidal intrusion front: an estuarine convergence zone. *Est. Coast. Shelf Sci.* **13**, 257–266.

Simpson JH and Turrell WR, 1986. Convergent fronts in the circulation of tidal estuaries. In: *Estuarine Variability* (Ed. DA Wolfe), Academic Press, Orlando, 139–152.

Simpson JH, Brown J, Matthews J and Allen G, 1990. Tidal straining, density currents, and stirring in the control of estuarine stratification. *Estuaries* **13**, 125–132.

Skreslet S and Loeng H, 1977. Deep water renewal and associated processes in Skjomen, a fjord in North Norway. *Est. Coast. Mar. Sci.* **5**, 383–398.

Slinger JH and Largier JL, 1990. The evolution of thermohaline structure in a closed estuary. *S. Afr. J. Aquat. Sci.* **16**, 60–77.

Smith R, 1980. Buoyancy effects upon longitudinal dispersion in wide well mixed estuaries. *Phil. Trans. Roy. Soc. London* **A296**, 467–496.

Speer PE and Aubrey DG, 1985. A study of non-linear tidal propagation in shallow inlet/estuarine systems Part II: theory. *Est. Coast. Shelf Sci.* **21**, 207–224.

Sternberg RW, 1968. Friction factors in tidal channels with differing bed roughness. *J. Mar. Geol.* **6**, 243–260.

Stigebrandt A, 1981. A mechanism governing the estuarine circulation in deep, stratified fjords. *Est. Coast. Shelf Sci.* **13**, 197–211.

Stommel H, 1953a. The role of density currents in estuaries. *Proc. Minn. Int. Hydrol. Conv.*, 305–312.

Stommel H, 1953b. Computation of pollution in a vertically mixed estuary. *Sewage Indust. Wastes* **25**, 1065–1071.

Stommel H and Farmer HG, 1953. Control of salinity in an estuary by a transition. *J. Mar. Res.* **12**, 13–20.

Sturley DR and Dyer KR, 1992. A topographically induced internal wave and mixing in the Tamar Estuary. In: *Dynamics and Exchanges in Estuaries and the Coastal Zone* (Ed. D Prandle), American Geophysical Union, Washington DC, 57–74.

Su J and Wang K, 1986. The suspended sediment balance in the Changjiang Estuary. *Est. Coast. Shelf Sci.* **23**, 81–98.

Taylor G, 1954. The dispersion of matter in turbulent flow through a pipe. *Proc. Roy. Soc. London* **A223**, 446–468.

Thorpe SA, 1969. Experiments on the stability of stratified shear flows. *Radio Sci.* **4**, 1327–1331.

Thorpe SA, 1973. Turbulence in stably stratified fluids: a review of laboratory experiments. *Boundary-Layer Meteorol.* **5**, 95–119.

Tully JP, 1949. Oceanography and prediction of pulp mill pollution in Alberni Inlet. *Bull. Fish. Res. Board Can.* **83**, 1–169.

Tully JP, 1958. On structure, entrainment and transport in estuarine embayments. *J. Mar. Res.* **17**, 523–535.

Turrell WR, Brown J and Simpson JH, 1996. Salt intrusion and secondary flow in a shallow, well-mixed estuary. *Est. Coast. Shelf Sci.* **42**, 153–169.

Unesco, 1987. *International Oceanographic Tables. UNESCO Technical Papers in Marine Science* **40**, Unesco, Paris, 195 pp.

Uncles RJ and Stephens JA, 1989. Distributions of suspended sediment at high water in a macrotidal estuary. *J. Geophys. Res.* **94**, 14 395–14 405.

Uncles RJ, Bale AJ, Howland RJM, Morris AW and Elliott RCA, 1983. Salinity of surface water in a partially-mixed estuary, and its dispersion at low run-off. *Oceanol. Acta* **6**, 289–296.

Uncles RJ, Elliott RCA and Weston SA, 1984. Lateral distributions of water, salt and sediment transport in a partly mixed estuary. In: *Proceedings of the 19th Coastal Engineering Conference, Houston*, ASCE, Washington DC, 3067–3077.

Uncles RJ, Elliott RCA and Weston SA, 1985a. Dispersion of salt and suspended sediment in a partly mixed estuary. *Estuaries* **8**, 256–269.

Uncles RJ, Elliott RCA and Weston SA, 1985b. Observed fluxes of water, salt and suspended sediment in a partly mixed estuary. *Est. Coast. Shelf Sci.* **20**, 147–167.

Uncles RJ, Stephens JA and Woodrow TY, 1988. Seasonal cycling of estuarine sediment and contaminant transport. *Estuaries* **11**, 108–116.

Van Dongeren A, 1992. A model of the morphological behaviour and stability of channels and flats in tidal basins. *Delft Hydraulics Rep.* **H824.55**.

Van de Kreeke J, 1988. Dispersion in shallow estuaries. In: *Hydrodynamics of Estuaries* (Ed. B Kjerfve), CRC Press, Boca Raton, 27–39.

Van de Kreeke J and Robaczewska K, 1989. Effect of wind on the vertical circulation and stratification in the Volkerak Estuary. *Neth. J. Sea Res.* **23**, 239–253.

van Straaten LMJU and Kuenen PLH, 1958. Tidal action as the cause of clay accumulation. *J. Sedim. Petrol.* **28**, 406–413.

Weisberg RH, 1976. A note on estuarine mean flow estimation. *J. Mar. Res.* **34**, 387–394.

West JR and Cotton AP, 1981. The measurement of diffusion coefficients in the Conwy Estuary. *Est. Coast. Shelf Sci.* **12**, 323–336.

West JR and Shiono K, 1985. A note on turbulent perturbations of salinity in a partially mixed estuary. *Est. Coast. Shelf Sci.* **20**, 55–78.

West JR and Shiono K, 1988. Vertical turbulent mixing processes on ebb tides in partially mixed estuaries. *Est. Coast. Shelf Sci.* **26**, 51–66.

West JR and Williams DJA, 1972. An evaluation of mixing in the Tay Estuary. In: *Proceedings of 13th Coastal Engineering Conference*, ASCE, Washington DC, 2153–2169.

Wood T, 1979. A modification of existing simple segmented tidal prism models of mixing in estuaries. *Est. Coast. Mar. Sci.* **8**, 339–347.

Wright LD, 1971. Hydrography of South Pass, Mississippi River. *Proc. Am. Soc. Civ. Eng.* **97**, WW3, 491–504.

Wright LD and Coleman JM, 1973. Variations in morphology of major river deltas as functions of ocean wave and river discharge regimes. *Am. Assoc. Petrol. Geol.* **57**, 370–398.

Wright LD, Coleman JM and Thom BG, 1973. Processes of channel development in a high-tide-range environment: Cambridge Gulf-Ord River delta, Western Australia. *J. Geol.* **81**, 15–41.

Yin K, Harrison PJ, Pond S and Beamish RJ, 1995. Entrainment of nitrate in the Fraser River Estuary and its biological implications. 1 Effects of the salt wedge. *Est. Coast. Shelf Sci.* **40**, 505–528.

Yoshida S, 1986. The growth mechanism of interfacial wave packet. *Bull. Fac. Engin. Hokkaido Univ.* **130**, 127–135.

Index